The Basics of
Regenerative Agriculture

ROSS MARS

Permanent Publications

Published by
Permanent Publications
Hyden House Ltd
13 Clovelly Road
Portsmouth
PO4 8DL
United Kingdom
Tel: 01730 776 582
Email: enquiries@permaculture.co.uk
Web: www.permanentpublications.co.uk

Distributed in North America by
Chelsea Green Publishing Company, PO Box 428, White River Junction, VT 05001, USA
www.chelseagreen.com

Distributed in Australia by
Peribo Pty Limited, 58 Beaumont Road, Mt Kuring-Gai, NSW 2080 Australia
https://peribo.com.au

Designed by Two Plus George Limited, info@twoplusgeorge.co.uk

Printed in the UK by Bell & Bain, Thornliebank, Glasgow

This product is made of material from well-managed FSC®-certified
forests and from recycled materials and other controlled sources.

The Forest Stewardship Council ® (FSC) is a non-profit international
organisation established to promote the responsible management of
the world's forests. Products carrying the FSC label are independently
certified to assure consumers that they come from forests that are
managed to meet the social, economic and ecological needs of
present and future generations.

British Library Cataloguing-in-Publication Data
A catalogue record for this book is available from the British Library

ISBN 978 1 85623 273 9

Praise for the book

Trying to capture both the fundamental essence and fringe nuances of regenerative agriculture in one volume is a Herculean task, but Ross Mars nails it in this fast-paced yet detailed book. From soil to family, he addresses the metes and bounds of a healing agriculture; what a joy to see this much heart compressed in such a readable, lively package.

Joel Salatin,
Polyface Farm and Editor of
The Stockman Grass Farmer

...Dr. Ross Mars has created an outstanding and timely summary of the emerging field of 'regenerative agriculture'. Outstanding as it captures and distils the wide body of practices involved; timely as regenerative agriculture becomes a buzz word and at risk of being co-opted to the point where it means little. Backing up on this widely-respected (and used) 1996 title The Basics of Permaculture Design, *Ross has used his skills as a practitioner and as a scholar to help anyone not only understand what regenerative agriculture is about, but also what are the key strategies, tools and practices available to manage any transition...*

Darren J. Doherty,
CPAg (AIA), Founder and
Director of Regrarians Ltd.

This artfully condensed read will help to set you on clear, nature-rich path towards abundance and homeostasis.

Nick Viney,
Regenerative farmer and rewilder,
www.keepitwild.uk

This timely book will enable the reader to better understand the important role that the permaculture philosophy and principles have exerted in influencing the development of organic, biodynamic and regenerative farming and food systems. It is quite literally a guide for the perplexed, throwing light on a fascinating chapter on the history of agriculture and food production.

Patrick Holden,
Founder and CEO of
Sustainable Food Trust

There are many regenerative agricultural practices demonstrating to farmers how to support and restore their land and maintain ecological and economic viability. Ross has written his latest book in an easy-to-understand language bringing us examples of regenerative agriculture and their principles together with the importance of soil and its care alongside the nuts and bolt topics of integrating plants and animals with the steps landholders may take transitional to regenerative agriculture. The Basics of Regenerative Agriculture *will talk to a wide audience from the discerning food buyer to the upcoming land steward, both driving the emergence of regenerative agricultural practices as the only way forward to heal and restore the health of our ecosystems and everything within.*

Fiona Blackham,
Gaia Permaculture

Regenerative agriculture comes in many forms and under many names. In this book, Ross Mars pulls together the different strands to show one coherent picture of regenerative agriculture and all that it entails. From rural to urban and from physical practices on the land to financial and social impacts of the way we produce our food, many different topics are included. He provides ethical as well as practical guidelines to consider and a scientific base from which to make decisions. I highly recommend this book to all farmers and food producers.

Martina Hoeppner,
Permaculture Education Alliance

Contents

To my wife Jenny, children, step-children,
grandchildren and great grandchildren,
who give my life meaning and purpose.

Foreword

Meaningful and life-changing learning can always be traced back to insights from deep engagement with nature and people, including teachers who teach from who they are, not just from what they know. Ross is such a teacher, who shares his essence and vast experience with agriculture, and especially the living soil on which it depends. This is a short, readable and useful book because it focuses on core wisdoms, understandings and practices, and does not include distractive 'fluff'!

Be ready for a regenerating feast – prepared by a master chef, who has been preparing to present this nutritious meal over a lifetime of foundational experiences, profound learnings and deep reflection. This is Ross Mars' crowning glory, in which he shares the what, why and how of regenerative farming – with the remaining questions and yet to be resolved issues included. Because of this, it is both a rare and valuable book, that I hope gets into the hands of all producers who aspire to be genuine regenerative agriculturists.

It takes the reader from the shallows and transitional stages of substitution regeneration (just based on alternative inputs) to the deep challenges of whole agroecosystem redesign.

Ross's recognition of healthy soil as the foundation of all of this is central to his story. I would only add that for this to be able to be achieved, radical personal transformation will also be essential. Such change will take courage, curiosity, experimentation, collaboration and persistence (what I call 'stickability'). Ross is a model of these qualities, and because of this he has been a mentor to so many throughout his teaching life.

In this valuable book he shares his understandings and wisdom. You the readers are the fortunate beneficiaries of these precious gifts.

Emeritus Professor Stuart B. Hill
Foundation Chair of Social Ecology, School of Education,
Western Sydney University (Kingswood Campus)

Preface

When putting this book together I did not accept or use information from popular articles or other writers without verifying their authenticity, nor did I want to write a scientific exposé quoting authors and findings to support my position. Rather, I used scientific papers to provide the factual information which is presented here.

The following people have kindly reviewed and commented on various chapters or offered advice and materials, and I am indebted for their support and encouragement:

Charles Massy (*Call of the Reed Warbler*)
Dick Richardson (Grazing Naturally)
Jeff Moyer (Rodale Institute)
Darren Doherty (Regrarians®)
Rachelle Armstrong (Soil Restoration Farming)
Dr. Christine Jones (Amazing Carbon)
Colin Seis (Pasture Cropping)
Mike Parish (Healthy Soils Australia)
Fiona Blackham (Gaia Permaculture)
Lorraine and Ethan Gordon (Southern Cross University, Regenerative Agriculture Alliance)
Martina Hoeppner (Dandelion Permaculture)
Ray Milidoni (Farming Secrets)
David and Frances Pollock (Wooleen Station)
Tomas Lewis – illustrations

1

How Do We Define
Regenerative Agriculture?

It is probably not a good start to this book to say that regenerative agriculture is hard to define. There is no single statement that seems to explain or define exactly what regenerative agriculture is, nor is it aligned to a particular discipline, although there are some common threads in the various definitions and explanations. What regenerative agriculture means depends on who you speak to or what books you read. Even books with regenerative agriculture in their title rarely discuss the meaning of the phrase.

The different ideas about regenerative agriculture are a reflection of the larger than life personalities of individuals beating their drum. Consensus amongst regenerative agriculture believers is difficult to obtain as each has their own spin of the 'correct' way of regenerative farming. There are many proven techniques and practices that do contribute to improved pastures and soil health etc., but many of these are driven by individuals, so the movement as a whole is fragmented.

Regenerative agriculture is just one of the terms associated with environmentally-friendly farming that also includes: Agroecology, sustainable, organic, permaculture, beyond organic, beyond sustainable, conservation agriculture, systems-based approach and restorative agriculture. These terms differ in their intent from conventional farming and all are seen as alternative agriculture.

Most consumers do not know much about agriculture, are disconnected from food production and most probably don't even know a farmer personally, but they seem to have opinions about how food should be grown. Current agriculture is a broken system but it can be fixed. We can't blame farmers for the past – we need them to feel that they are part of the solution, and we need to encourage farmers to stay on the land.

What we call conventional or traditional farming today, as a description of what farming practices we have adopted worldwide after World War II and the advent of chemical fertilisers, herbicides and pesticides, implies that this is normal farming as opposed to say organic farming which doesn't use these types of chemicals.

After WWII there was a huge push to use all of the fertilisers and pesticides that were developed during those times. The Green Revolution was born when new machinery replaced horses and made agriculture easier. Unfortunately, while the Green Revolution initially gave the world promise, environmental degradation soon started to happen, and now it is critical that the way we farm and grow food has to change. It also became evident in more recent years that we can't rely on technology and genetic manipulation to solve the food problems of the world.

'Conventional' is a misnomer as it implies normal, acceptable, commonplace. By conventional we really mean industrial, chemical-driven farming. Most farming practices today are akin to mining. The soil is excavated, nutrients are removed from the soil, the nutrients (bound in crops) are hauled away. Most mining operations have an obligation to fill the hole in and make good by replanting natural, local vegetation. Farmers also need to replace some nutrients by way of fertilisers and save seed of the exotic crop and this cycle of remineralisation and seed-saving has to repeat itself year after year.

So industrial agriculture is more about extraction than preservation, more about controlling nature than working with nature, more about monocultures than polycultures and more about profits than ecological accountability. Unfortunately, there is a deeply embedded belief that humans should subdue nature, dominate it. The ingrained culture of anthropocentric (human-centred) ideology has been slow to change. Humans must become more aware of their place in nature rather than seeing themselves as masters over nature.

The industrial agriculture model only concerns itself with inputs, outputs and profit. With industrial agriculture it's like growing crops on steroids. There is no consideration for soil improvement, environmental health, water quality, nutrient cycles, water infiltration, erosion and sediment control, and soil life. In some countries government assistance props up the industrial farming enterprise. Industrial agriculture is a flawed system. It promises much but delivers little.

Conventional farmers rely on high inputs and they need to meet certain specifications in their produce – which could be size, colour, shape, amount of protein and weight. This then puts stress on the farmer to meet these standards or their produce cannot be accepted.

Figure 1.1 Industrial agriculture compared to regenerative agriculture.

When the cost of inputs increases, the farmer may not be able to get a higher commodity price. This translates to lower net income, and the only choice a farmer may have is to pay less for labour or to reduce other expenses.

Organic farming was initially traditional farming but is seen by many as expensive food. The moral issue is how to make organic food cheap enough for everyone, not just the wealthy who can afford premium prices.

Many low income families do not have the luxury of choice, and individuals who have poor education and are not informed about food quality, don't see organic food as an option either. Compounding this is that industrial farmers are often powerless to change, even if they want to. They are locked into repayment contracts for machinery, seed, fertilisers, herbicides and pesticides. Conventional food production is often subsidised so its true cost would be similar or even higher, in some cases, than organic production.

The main difference between organic and conventional farming is that conventional farming relies on chemical intervention to provide plant nutrition and to control pests and weeds, whereas organic farming relies on other, more natural, techniques to minimise pest attack and improve soil fertility. However, there is more to this. When we compare soil conservation and fertility, water retention, the effect on the environment and protection of bushland and forests, water and air quality, and energy use, it is clear that organic farming is more responsible. If a farm is not being managed as an organic enterprise it is still industrial. This is because farmers still rely on agrichemicals to some degree. The table below is a simple comparison between the two.

Table 1.1 The comparison between conventional and organic farming

	Conventional	Organic
Yield	Assume to be the same	
Inputs	High	Low
Energy consumption	More	Less
Greenhouse gas production	High	Low
Water infiltration	Low	High
Soil health	Low	High
Profit	Low	High
Natural pest and disease resistance	Low	High

The term regenerative agriculture was first coined by Robert Rodale in the 1980s to distinguish a kind of farming that goes beyond sustainable. Sustainable agriculture was fashionable at one stage, but sustainable just meant maintaining what you have, and not so much improving for the future.

The meaning of sustainable can be misleading. It implies being stable and keeping the same, so sustainable agriculture might suggest keeping the same or at least no further deterioration. But, of course, the agriculture being practiced might not be that great, so sustainable in this sense is keeping or promoting these terrible practices. We need to be beyond organic and beyond sustainable, which is regenerative.

When we first heard the term 'sustainable' it meant improving the environment without any further harm, but as the years rolled by it became a catchcry for anyone who promoted the message of 'staying the same' and in this context could be taken to mean sustaining a degraded ecosystem and not letting it get worse. Sustainable is about maintaining the status quo, it is not about continuous improvement and restoration of degraded landscapes and better food quality. Certainly, the common use of 'sustainability' is not seriously improving things which is what we require today if we want a better world for all.

Rodale promoted greater biodiversity in both the soil and farming operation, planting more perennials and reducing inputs by composting and making better use of local and on-farm resources. What springs to mind when you promote the concept of 'regenerative' is ethical integrity. There is a moral imperative to fix the problems we find ourselves in. Regenerative is promoted as being beyond sustainable.

Organic techniques, as promoted by the Rodale Institute over the last 50 years have always included cover crops, crop rotation, low or no till, and using compost and other soil amendments, all with the aim of improving the soil and increasing biodiversity on the farm.

It is a misnomer that organic food cannot feed the world – a myth perpetuated by big agribusiness and Earl Butz during his term as Secretary of Agriculture in the USA in the 1970s. Many studies have shown organic production and nutritional value to be the same or better than conventional agriculture. Various published papers suggest conventional production is better but some trials are biased. If the organic food was grown in good soil and not a degraded landscape, which is depleted of soil life and nutrients, then the results would speak for themselves.

Over the years there has been criticism that organic farming is not as productive as conventional farming and it won't be able to feed the world in the years to come. While there can be some truth in that idea we need to recognise that what we currently do is not working, is not sustainable, and chemical-led production will eventually cause soil collapse. Organic and regenerative agriculture allows us to produce year after year without soil deterioration. As soil health improves so does yield.

Thus, we shouldn't view organic production as always lower than conventional production because clearly that is not always the case. And as we develop a greater understanding of plants and microorganisms we will be able to enhance and support these relationships which ultimately result in quality food production. Over the next century it will be very apparent that the resilience of organic growing will come to the fore, and conventional agriculture will just be a distant memory.

In the early 1990s permaculture practitioners embraced the regenerative agriculture philosophy and applied the various permaculture principles to designing functional, integrated farms that built on water harvesting and movement, polyculture, intercropping and agroforestry, soil amelioration, sound stock management, perennial food crops and organic production. The mantra was "work with nature, not against it". At the end of the day, we cannot beat nature. We cannot improve on nature, and our actions will not lead to a more sophisticated level of nature.

Agriculture needs to move in the direction in which nature originally designed and we must not continue to view ourselves as being outside of nature. At times it seems nature is more collaborative than competitive than what we were led to believe and what we once thought. There are a lot more organisms working together and helping each other, than fierce competition where only the strong survive. Besides, if you give nature half a chance it will correct any problems.

Up to this point, the permaculture designers were mainly focussing on urban and small acreage rural properties. Now they could apply the design principles and strategies on a much larger scale.

Figure 1.2 Agroforestry and perennial food crops can be very productive systems.

At about the same time Alan Savory formulated his Holistic Management philosophy and provided a decision-making framework and a set of planning practices (such as grazing, financial and environmental) to better manage stock on the farm, and the focus was an animal-centred farming operation that worked to improve soil quality. Many holistic management farmers call their farming enterprises regenerative agriculture.

Regenerative agriculture is not a new way of farming. It could be argued that many Indigenous peoples all over the world farmed in ways to protect the soil and the local environment. They held a deep connection to the land and recognised the importance of preserving the land for future generations.

Rodale just put a name to these types of practices, a name that resonates with the slowly changing mindset of the farming community. Permaculture, in turn, adopted the techniques and practices of Indigenous peoples from many parts of the world, and it provided a framework to design resilient systems in the face of adversity.

A large number of regenerative farmers were once traditional or conventional farmers. They want to reduce chemical inputs, but not necessarily adopt full organic practices. There was a compromise: Endeavour to build soil but reduce their use of pesticides and herbicides. This appeals to the majority of farmers today, and along the way it also has appealed to climate change activists and others who want change to agricultural practices. This is where the movement is at.

Regenerative agriculture has the potential to provide healthy, nutrient-dense food, revitalise rural communities and improve local economies as well as combating climate change, reducing greenhouse gas emissions, protecting ecosystems and preserving local and traditional knowledge.

Regenerative agriculture has been defined by others as having some combination of any of the following: Improving soil health, increasing yield and productivity, carbon sequestration, producing healthier food, reversing climate change, building resilience in the agricultural system and viewing the farming landscape from an ecological perspective. In this context, agricultural problems are ecological problems.

Climate change is about climate vulnerability, experiencing more extremes in temperature, storms, winds and rainfall as well as the inevitable rising earth temperature and drying climate in some areas and proportionally more rainfall in other areas of the globe. Climate change can also be called climate uncertainty. Our planet is heating up, localised weather patterns are changing and the prospect of our climate being predictable is not that certain. Unfortunately, the action for climate change needs to be more rapid than the speed at which policies are changing.

Agriculture will clearly be affected by climate change. If the climate changes, much of agriculture is affected first, so there is a sense of urgency. However, agriculture also needs regenerative practices to make it become more resilient to both environmental and economic impacts of climate change. We must use natural solutions to draw carbon out of the atmosphere. Extensive field experiments all across the world have demonstrated the capability of regenerative farming practices to increase soil carbon. However, agriculture is not the only solution needed. There isn't one answer or one solution to climate change. We must address the energy, transportation and building impacts, as well as agriculture, to help solve the climate crisis.

The term regenerative is often associated with rejuvenating, or regenerating, soil that has so often been degraded by industrial farming since the 1950s. Regeneration is seen as the process of renewal, rebuilding, restoration and growth that makes organisms and ecosystems resilient to disturbance or change.

So, back to that definition. Some would define regenerative agriculture as a system of farming principles and practices built upon a soil foundation that uses carbon via photosynthesis to improve nutrient and water use efficiencies and provide resilience against climatic uncertainty and pest and disease issues.

Others might define regenerative agriculture as describing farming and grazing practices that, among other benefits, reverse climate change by rebuilding soil organic matter and restoring degraded soil biodiversity – resulting in both carbon drawdown and improving the water cycle.

Some definitions focus solely on soil health: Regenerative agriculture is any approach to farming that ensures that the soil improves over time. Others see regenerative agriculture as having no specific rules or practices. Some see grazing animals as the cornerstone of the regenerative agricultural movement. But I don't agree. While I am all for innovation and experimentation there are some things that are non-negotiable. It can't be 'do anything you think is helping'. As Mark Knopfler (Dire Straits) once sang "two men say they're Jesus, one of them must be wrong". There are some truths out there and these just have to be accepted.

Figure 1.3 A farm can be seen as an agricultural ecosystem.

Whatever definition of regenerative agriculture we adopt, it needs to be clear that we should not focus solely on practices but on outcomes: On what we hope to achieve. Outcomes-based definitions might involve some combination of improvements in people and land, increases in landscape function (how much litter, soil cover and soil life you might find) and biodiversity, and that any farm profit is at least stable and increasing at best. To clarify these ideas, practices might include holistic planned grazing, pasture cropping, natural sequence farming, keyline cultivation, whereas the outcomes we want to achieve could include increasing soil cover, increasing biodiversity and so on. These outcomes are morphed into various principles which are covered in the next chapter and various practices are outlined later in the book.

What is clear is that regenerative agriculture is a multifaceted approach where various principles and practices enhance and restore all components of an ecosystem. When this definition is expanded and fleshed out, all components of an ecosystem include humans (social capital, community), plants and animals, air and water quality, soil and all other aspects of the environment. We need to work towards growing soil, improving air and water quality to their optimum, and providing all resources to enable ecosystem function and literacy.

The regenerative agriculture food production model is not complicated, it is just beyond what most farmers know and accept. What farmers hold as their worldview, what farmers learnt from their parents as they grew up and what gets traditionally taught in agricultural colleges and universities is only a small part of what is possible. So, regenerative agriculture is more than improving the soil and putting in cover crops. What needs to shift is the paradigm that we hold from 'we have always done it that way' to 'what can I do differently that will make a difference?'. We need a fundamental shift in thinking about what we grow and how we grow it. Regenerative agriculture endeavours to consider everything, which includes environmental impact, profitability, increasing social capital and developing strategies to build resilience.

The issue is that there are no rules about regenerative agriculture. No person or organisation has set the rules that farmers should follow. At best we accept the principles on which regenerative agriculture is based and then we use whatever practices, technologies and resources are at our disposal.

Regenerative agriculture has much to recommend it as an ecologically sound alternative to industrial farming. It supports diverse crop rotations, better biological diversity such as habitat for pollinators, and healthy soils, which all benefit the environment. It may also contribute to higher incomes for struggling farmers. If the soil is improving, the input costs are diminishing, and the crop yields are maintained then common sense dictates a profit can be made.

Regenerative agriculture includes far more than just livestock, and utilises a broad range of conservation practices that include streamline vegetation and buffer strips to protect water quality, shelter belts and hedgerows to reduce wind erosion, and returning marginal lands to perennial plant cover. It's a different ethos, one that seeks to respect and mimic natural cycles and processes, and increases rather than decreases soil fertility over time.

Many farmers say that they are trying to mimic natural systems and they have varied success with this, but nothing about farming is natural. Farming is directed by humans in an agricultural ecosystem where inputs, outputs, matter cycling and energy flows are all quite different to a natural forest, bush or aquatic ecosystem. We farmed with pure ignorance of landscape function, with no understanding of the complexity of soil and fragile environments, and we were driven by greed. We farmed by focusing on yield, visual appearance and cost, mainly because consumers demanded it, whereas we should develop different criteria where sustainable food production was at the fore.

There is no recipe for a regenerative farming system. There is no one size fits all. Every farm is unique and requires specific management. While some of the common practices work universally, there may be some tweaking to ensure the practices are right for you.

Is there a generic model of regenerative agriculture to fit most farms? No. What gets adopted (and what is successful) varies from farm to farm, region to region, country to country. However, the basic principles to build a regenerative agricultural practice are common, and some of these are more relevant for some people, and others might embrace only some of the principles. These principles are applicable everywhere, in every soil type, in any climate and in any landscape. Nature is everywhere so regenerative agriculture can be everywhere too. Regenerative agriculture is complex, just like nature.

As already mentioned, the transition to regenerative agriculture is a paradigm shift, a shift in thinking about how land is used, food is grown and ecological literacy. Ecological literacy (or ecoliteracy) describes our understanding about living systems (ecosystems) and nature which make all life possible. It challenges our world view and what we hold as true. For example, it seems completely irresponsible if we continue to grow plants to feed animals or to supply industrial products such as ethanol to fuel vehicles. With potentially unlimited opportunities comes responsibility, and everyone is accountable.

Even though a farmer may feel trapped in an industrial system, a bad system, there comes a time when you have to take responsibility and there is no excuse for inaction. Scientific research continues to reveal the damaging effects to soil from tillage, applications of agricultural chemicals and soluble fertilisers, and reductions in soil biota. We know that conventional agriculture has caused many other problems such as water logging and salinity issues, erosion and loss of topsoil, insect, weed and pest plagues, and over-clearing of land and reduction in local wildlife habitats.

Regenerative agriculture has the opportunity to reverse soil destruction and to improve ecosystem function. The research we need to undertake often comes under the umbrella of agroecology. Agroecology is an applied science that studies ecological processes applied to agricultural production systems, but in the context of society, farming innovation and community values. Agroecology is independent from organic, regenerative, permaculture, holistic management, conventional farming and other disciplines, and is concerned with the interaction of humans with farming systems and the surrounding environment.

Agroecology is a multidisciplinary approach that analyses the impact of agricultural practices on the soil and environment as well as within a local (typically rural or regional) community, plant and animal interrelationships, economic constraints for farming enterprises and providing information about productivity and system stability.

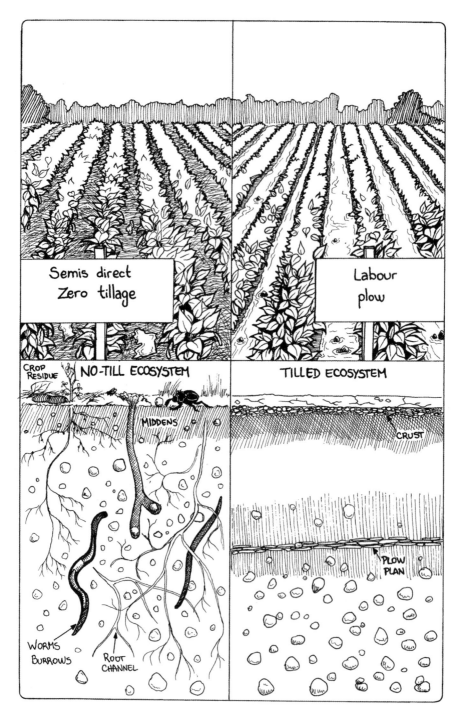

Figure 1.4 Tillage destroys the soil and
changes its structure and water permeability.

Modern agriculture is a right-livelihood occupation. Regenerative agriculture should be seen as a way to improve people's lives. It can bring personal fulfilment, increasing your sense of worth and a good way to make a living. But you need to feel a connection to country – to the land – if you truly believe that farming should be about good practice.

Every farm is different, every farmer's journey their own. There is no 'this is the way', dogma where regenerative agriculture practices suit everyone. You just need to see what works for you. If you follow the regenerative agriculture principles then success is inevitable.

Although we have focussed so far on rural farming you could ask the question: "Can regenerative agriculture occur in an urban environment?". It already is! There are city farms, community gardens, backyard production, farmers' markets, edible streetscapes, rooftop gardens and green walls. The agriculture doesn't have to involve beef cattle and other large animals, but chickens, ducks, geese, rabbits and bees are amongst this mix of urban production. The principles and techniques of regenerative agriculture should not be restricted to rural enterprises; urban agriculture also cries out for help.

Our future depends on a healthy environment. Food and farming are fundamental to this. In particular, we should focus on:

- No 'icides' – fungicides, pesticides, insecticides and herbicides

- Fertility – use natural soil amendments to help you get that soil fertility balance right; amendments are covered in the soil chapter

- Organisms that are essential for plants to thrive and survive

- Profitability – at the end of the day a farm has to be profitable and economically viable

- Enhancing the environment – no pollution of air, soil and water, no waste, protect waterways; leave a future for our children and future generations

- Building a farming ecosystem – enhancing biodiversity

- Valuing community and human interaction

- Social justice – embracing the pillars of sustainability, encouraging right livelihood occupations

- Securing future food sovereignty, to ensure our physical and mental health and well-being

- Meeting the needs of all including all elements – plants, animals, humans

- Designing for resilience, designing for catastrophe

- Enhancing nutrient cycles.

One way to have a general overview of what regenerative agriculture is all about is to examine the simple acrostic (word puzzle) below. An acrostic describes terms or concepts for each letter of a word. So, each letter in regenerative agriculture has a word or phrase to help explain its meaning.

Resilience
Energy
Guilds, **G**roundcover
Economics
No till nutrients
Ecosystem, **E**cology
Recycling
Air
Thriving business
Integrated, **I**ncrease organic matter
Vegetation
Environment, **E**cological literacy

Amendments, **A**nimals
Grow soil
Recovery
Increase water absorption
Culture, **C**rops
Underground biota
Landscape design
Terrestrial
Unicellular organisms
Right livelihood
Encourage biodiversity

The regenerative agriculture acrostic is something I made up. I am sure you could think of other words to help explain regenerative agriculture and what it entails. However, it should be clear by now that we need to move from small, single conservation practices to a more holistic one involving the whole community. We shouldn't shame farmers into change but encourage them to explore different options. We need to show them that they can farm differently and still be profitable. We all need to become biochemical and biological farmers.

2

The Principles of Regenerative Agriculture

Humans are not separate from nature. We need to acknowledge our place in the living world and, ultimately, that everything that affects nature will eventually affect us. Most thoughts about regenerative agriculture list various principles, mainly focussing on improving the soil. While all of these, in one way or another, form part of the following list, a truly regenerative system also needs to consider the impact of any farming system on the environment, on the local community, on all organisms on the farm and the surrounding area, on the natural processes and cycles of matter, as well as developing strategies to build a resilient system that can withstand predicted changes to our climate, fluctuation in pests and disease, and, with all of this, still be profitable.

These principles provide some guidelines to consider as your journey into regenerative agriculture starts or continues. They are not in any order of importance and, for you, some may have more relevance than others, but any regenerative enterprise should keep these overarching principles in mind.

We could argue what is a principle and what is a practice. Is 'cover the soil' a principle or a practice? I tend to think of practices as the techniques we undertake on our farming properties. For example, we might say "I practice holistic planned grazing" or "I have set up a permaculture property" or "I follow the Regrarians Platform®".

However, Ethan and Lorraine Gordon (Southern Cross University) have proposed a set of more philosophical principles which are an overarching framework that can be used to make decisions about what actions and practices would be suitable for a particular farm. Their principles are briefly:

1. Think holistically: Farms and farming should be seen as a 'whole' so all aspects of this and the environment are considered.

2. Have an understanding of complex adaptive systems: We need to have a good understanding of nature as both complex and dynamic.

3. Be comfortable in ambiguity: We need to accept that we cannot control or know everything, so endeavour to do the best you can with what you have.

4. Have the capacity for continuous, transformative learning: We need to reflect on our world view and recognise and embrace that this can alter as we learn more about our place in nature.

5. Understand that human cultures are co-evolving with their environments: Our biological and cultural roles within landscapes change along with any environmental and ecological community changes.

6. Make place-based decisions within bioregions: Any changes to farming practices should be site-specific. Some practices will be inappropriate in some areas and at some times.

7. Acknowledge and involve diverse ways of knowing and being in landscapes: Consider various ecological perspectives and use those that align with your vision.

To me, principles are those beliefs guiding our actions, so 'cover the soil' is an overarching statement of what we need to do. How that is achieved is the practice we employ: Use mulch, plant a cover crop or plant a perennial pasture. If we are transparent and honest, then healthy discussion will follow. The following 20 principles can be seen as complementary and inclusive of those philosophical ones developed by Southern Cross University.

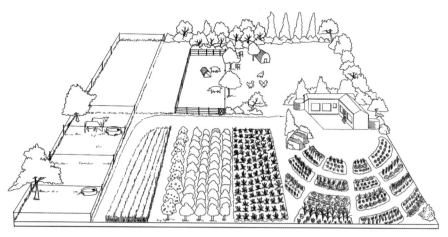

Figure 2.1 A whole farm plan should be developed to showcase the vision you have and guide the implementation.

1. Know your Farm

This probably is the most important principle. Farmers must know all aspects of their operation, including the various soil types present, the nutrient status of these soils, annual and monthly rainfall patterns, prevailing winds in each season, sun angles and sun movement during summer and winter to enable them to determine areas of shade in the morning or afternoon, frost lines, water movement across the landscape, typical seasonal temperature ranges, areas that tend to become waterlogged in winter, and water quality of their dams, soaks, bores or creeks.

Some of these observations will help inform whether various microclimates can be exploited to increase production. For example, you should note which areas are shaded longer as they may be on the south (or north) slope or alongside large stands of trees and the sun doesn't hit the ground till late morning or vice versa – the trees shade the other side of the paddock from mid-afternoon onwards. All of these observations and the climate and landscape data are essential to develop a whole farm plan.

This data enables you to have a baseline, a starting point, from which you can measure improvements or deterioration, as well as providing the information you need to develop the plan and way forward. You will also need contour maps of the property to help plan new dams and water capture and movement.

By compiling the data for your property you will have a good picture of when rains first come or are last to arrive, the maximum and minimum temperatures each day so you know what months or parts of months typically bring frost, how your soil is changing as you improve fertility or add amendments to correct out of whack soil acidity or alkalinity, and soil temperature and moisture when and if the need arises as you begin your journey of improvement. You need a benchmark so that you know what you are starting with and can measure future improvements.

2. Design for Resilience

Many farmers and organisations working in the regenerative agriculture space seem to focus on climate change. While this is crucial, designing for resilience asks you to also consider future market forces, potential crops more suitable in a changing climate, ways to improve soil fertility, environmental enhancement, social capital, and working out strategies to enable you and your family to strive and survive in a rapidly changing world. We will all need that mental toughness as the future unfolds.

Designing also implies a plan, like a whole farm plan with details about species, timeline for implementation, resources available and needed, estimated budget for various projects and careful consideration of the spatial placement of new structures, earthworks and multifunctional plant and animal systems.

The design process itself may not be a simple linear one as ecosystems are complex, although it can usually be done in a series of steps. Each farm design is site-specific and context-based, but connections between organisms are recognised and drawings evolve about the eventual layout of all of the components in the farm. In any case, think about a design in nature, with nature in mind.

3. Meet the Needs of All – Respect all Life

We need to recognise and understand that all living things are worthwhile and have their particular roles or niches in the environment, although I must confess, I struggle to understand why we have mosquitoes, ticks and viruses.

We need to assess the uses, function and products, wherever applicable, of whatever we have in our farming systems to see how each of the components of the system can be integrated, so that the needs of one are met by the products, services or functions of another. For example, planting a nitrogen-fixing cover crop provides many products, uses and functions. The ground cover protects the soil, the exudates (substances released from plants) from its roots feed beneficial fungi and microbes, the plant may be grazed or mown to provide food for stock or more organic matter for the soil, nitrogen will be released either when the plant dies away or via the manure that stock excretes thus improving soil fertility, the above-ground plant mass can be cut and made into hay or silage, and soil structure will improve as roots penetrate deeper into it. As it is in every natural ecosystem, the plants, animals and microorganisms found there all support each other in various ways.

Finally, you must ensure that all of the needs you and your family members have are met, as they are the most important parts of the farming operation.

4. Integrate all Components of the Farm

If we acknowledge that we want to have an agricultural ecosystem on our farms then the question becomes: 'How do we integrate all of the components (what permaculture practitioners call elements) so that the wastes or products of one become the needs of, and used by, another. That plants requiring pollination can have this ecosystem service performed by particular insects or other organisms, and if animals need protection from harsh weather, what shelterbelts and windbreak tree and shrub species will need to be planted?'

We shouldn't just think about what trees, shrubs, ground covers, crops and animals we want to have or grow, but also the other elements we need to support these. These other components include plants we use to attract predators for pest control, fodder species for stock, a particular mix of cool season and warm season cover crops, native (local endemic) plants to rehabilitate degraded landscapes, plants to entice pollinators and amendments including manures to build soil fertility and health.

Ideally, we look for multifunctional elements – those components that provide several products or functions for the system. For example, wattles such as *Acacia* spp. (Australia), Italian alder *Alnus cordata* and sea buckthorn *Hippophae rhamnoides* (Europe) or black locust *Robinia pseudoacacia* (America) are all nitrogen-fixing trees that are fast-growing pioneer plants that may have some combination of uses such as fodder, windbreaks, providing flowers for pollinators and for honey production, fruit or seed for animals to eat, timber or firewood, predator attracting, grows in a range of soils, may have some salt tolerance, either evergreen or deciduous, or acts as a nurse plant to protect other trees.

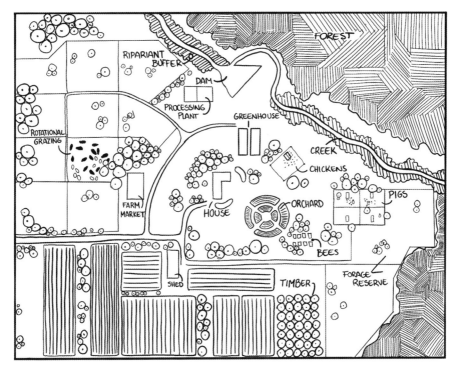

Figure 2.2 Design for the integration of all components on a farm.

5. Maximise Biodiversity

Every natural ecosystem exhibits high plant and animal diversity. It is true that a greater diversity above the soil equates to a greater diversity below the soil. We need to develop polycultures on our farms and move away from a single crop (monocultures of large areas of wheat, corn, soybeans or barley, for example). Not only should we aim for crop diversity, but if we plant a cover crop for animal fodder then we shouldn't restrict this to only a few species. The more biodiversity we have on our farms the more robust and resilient it will be. Diversity helps with the stability or resilience in nature while monocultures are vulnerable. When we look at nature we never see monocultures of particular plants.

Building biological ecosystem diversity is a mirror to what happens in nature. Biodiversity is the key to resilience when catastrophes affect an ecosystem. There is enough variation in plant populations to withstand external pressures and to bounce back, enough groundcover to protect the soil and soil life and enough species (multispecies) to continue with essential ecosystem functions (including pollination, predation and nutrient cycling). Having more diversification on our land ensures a successful farming operation.

6. Build Healthy Systems

Healthy farming systems include soil fertility and soil health, human health and well-being as well as plant and animal health. We only have to look at the decline in human health over the last few generations, the increase in lifestyle diseases and the nutrient loss in vegetables over time – where modern foods are less nutrient dense and lower in minerals and vitamins – to realise that we can do better. We can be healthier, fitter and live more fulfilling lives.

We all love to grow plants, but what we should focus on is that we need to grow soil. That is as simple as the message has to be. When we improve the health of our soils we improve all other aspects on the farm.

We also need to produce more nutritionally dense food, and this can be difficult when the area of arable land is decreasing every year. As our agricultural lands become depleted and degraded, the pressure on us is to clear more forest and bush to farm. This is counter-productive if we all believe that we should be planting trees, among lots of other things, to combat climate change. We just need to hold on to what has been cleared and work with that to make it productive.

Soil fertility can be increased through a range of biological procedures and practices which include cover crops, crop rotations, adding compost, compost extracts and other soil amendments, and animal manures in conjunction with increasing soil biota.

Growing food is not that hard, but we need to eat better (eating whole foods, not processed foods) and reduce our food miles – recognising that much of what we eat comes from overseas or interstate.

7. Increase Organic Carbon in the Soil

We need to look at the long term rather than the quick gains in the short term. There is more carbon below ground than found in the atmosphere and above ground combined. All we have been doing over the last few decades is removing that stored carbon when we extract and use fossil fuels and when we cultivate the soil and engage in conventional agricultural practices (such as burning stubble and tillage that destroys soil structure).

Carbon farming seems to be flavour of the month, and rightly so. All life depends on carbon which is the universal key to manufacturing organic substances that are found in plants and animals. The carbon cycle drives all of the other nutrient cycles in one way or another. We don't need much carbon or organic matter in our soils to grow food. We can get away with 5% as an indicative level of organic matter, but unfortunately many farming soils are less than 2% carbon.

Farmers need to increase and maintain reasonable levels of organic matter to provide the raw materials that sustain all soil life. This can be achieved by planting cover crops, chop and drop plant material onto the ground, adding biofertilisers, composts and amendments to soils, and introducing beneficial microbe inoculants to do their natural work to break down organic matter and make it more readily available to plants.

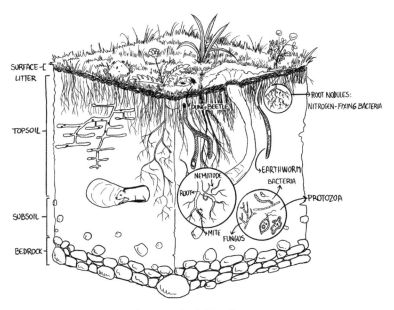

Figure 2.3 Soil is alive.

8. Minimise Soil Disturbance and Keep the Soil Covered

When we disturb the soil by turning it over or simply breaking the surface, we kill organisms, change soil structure and cause nature to respond by allowing weeds to grow. Tillage breaks up the soil surface and often turns the soil over. The aim of tillage is simply to kill weeds and provide furrows and cracks where seed can be sown. However, it also causes carbon loss, soil erosion, death of microbes including fungi and changes to soil structure. Over-working the land by cultivation only leads to compaction, reduced biology, low levels of carbon and organic matter and lower water retention. On the other hand, no-till enables the opposite to occur: Carbon sequestration, reduced soil erosion, growth of soil life and better soil structure and fertility.

While ideally using less tillage, it is a bit hard if you grow vegetables such as carrots, potatoes, Jerusalem artichokes and turnips, so tillage practices do depend on the crop that is grown.

Figure 2.4 (overleaf) shows the importance of soil cover. The left diagram shows soil temperature over the course of a day and it is clear that soils with cover have much lower temperature changes. The right diagram shows how soil moisture depletes as the temperature rises. Again, the temperature of bare soil rapidly falls and less change is noted for a pasture cover. This equates to less water loss occurring if the soil is covered by plants.

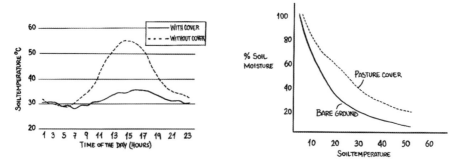

Figure 2.4 Soil moisture changes with temperature.

Minimal or no soil disturbance may be difficult to achieve, simply because conventional farmers have dedicated machinery and have continued with practices they inherited from European farming techniques. Although no or minimal till is fast becoming adopted worldwide, it is often associated with increased herbicide use. And this is the dilemma. How can we not use herbicides to reduce weed infestation to enable our crops to grow and not be contaminated by weed seed?

Furthermore, keyline ploughing has been shown to be beneficial to allow more water into the soil and to enable the soil to hold it higher in the landscape. So, on one hand we aim for essentially no-till and on the other hand we want more water held in our soils. Let's examine all of this a little later.

It is very rare to find bare soil in a forest or bushland. Even if there is only one centimetre of soil on top of rock, there will be plants of some sort. Nature wants soil covered at all times. Nature never tills the ground, other than certain animals that burrow in search of food, and protect it from desiccation and abrasion so that the soil life has every chance of survival.

Maintaining plant cover on soil also enables a living root maze all year round. The plant roots are where the action is. They are the interface with soil particles, nutrients and soil organisms, where substances are exchanged, water and gases absorbed, humus is formed and where organic materials are stored and released. This is why stubble retention is important if we cut a crop. The stubble needs to be retained, ideally flattened so it can break down faster, but certainly not burnt. Burning stubble results in the rapid loss of carbon, as well as high losses of nitrogen, phosphorus, sulphur and potassium. This practice must stop.

9. Farm Organically

Even though regenerative agriculture arose out of the organic farming movement it has been adopted by conventional farmers who continue to use pesticides as part of the mix and generally have chemical-dependent operations. As farmers transition to more regenerative agriculture practices it is hoped that reliance on herbicides and pesticides will decrease and eventually stop. Some of these chemicals have been shown to damage soil life which drives carbon sequestration and soil fertility, as well as the risk of chemical exposure increasing diseases in both children and adults.

Regenerative farming focuses on restoring soils that have been degraded by industrial agriculture and ultimately relies on nature doing the work. However, for some farmers this transition needs to be gradual as they endeavour to reduce their reliance on artificial chemicals.

Even so, we cannot ignore the elephant in the room. Unfortunately, most farmers are dependent on chemicals to control weeds and pests, and it is hard for them to wean themselves off their use. Many regenerative agriculture farmers recognise that they need to cut down and they have drastically reduced chemical use, but going organic is another leap.

Truly regenerative agriculture has to be organic. It shouldn't use artificial fertilisers and pesticides, simply because of the dangers of using these chemicals and their impact on beneficial organisms, soil life and the environment itself. We need to get the farmers to transition from a chemical dependent operation to a biological operation. We need to gently get farmers to change and bring all of those people through a paradigm shift. The task won't be easy.

We all acknowledge that there is ample evidence that fertilisers work. Crop yields also increase and more nutrients in grass equates to higher animal production. But using artificial, water-soluble fertilisers is only a quick fix and temporary solution. You have to apply them every year because when the crops are harvested the nutrients you applied last year or during that growing season went out through the farm gate as produce.

10. Increase Soil Life

The soil and soil life is key to successful agriculture, and every effort needs to be made to enhance these. Although estimates vary, it has been suggested that there are more organisms (mainly bacteria) in every heaped tablespoon of healthy soil than all of the people living on Earth – that's now over 8 billion (January 2023). With about 12% (1.5 billion hectares) of the earth's surface devoted to crop production, that's a lot of life, especially when there is about 1.5kg (3.3lbs) of organisms for every square metre of soil (for a 200mm depth). It is hard to comprehend, but basically there is as much biodiversity below ground as above ground, although the total biomass varies with ecosystems. For example, there is more biomass below the soil surface than on top of the soil in many grasslands, but the opposite is true for forests, where there could be four times more biomass above ground.

11. Store More Water in the Soil

Without water there is no life, it is the driver for all cellular reactions including photosynthesis. Whenever water is present in a soil, life flourishes. Unfortunately, various soils hold different amounts. For example, sandy soil may only hold up to 1mm/cm (0.5-1in/ft), loams 1 to 3mm/cm (1-3in/ft) while clays

2-5mm/cm (2.5-4in/ft). However, while clays can hold more water it is not available to plants because the clay particles strongly hold onto the water molecules and won't release them. Loam holds a fair amount of water, with proportionally more of this available, so this is best for growing plants.

Not only do we need to change our soils into more loam-like, we need to find ways in which more water can be held. Whatever rain that falls should enter the ground, with ideally no run-off which could cause soil erosion and loss of arable land. For farming, water is the key resource under pressure. While water vapour is a greenhouse gas and high production of water through industrial reactions and processes is adding to the greenhouse phenomenon, managing water is crucial for farm production.

12. Enable Nutrient Cycling

There is no waste in nature. Everything eventually breaks down and is reused by organisms. While humans do produce non-biodegradable waste, we need to do better. Every ecosystem has its nutrient recyclers, which are mainly small organisms such as dung beetles and black soldier flies to microscopic fungi and bacteria. They break down complex organic matter and substances into simpler parts, and these can be re-absorbed or used by other animals as food sources or by plants to make new substances such as sugars, starches and protein.

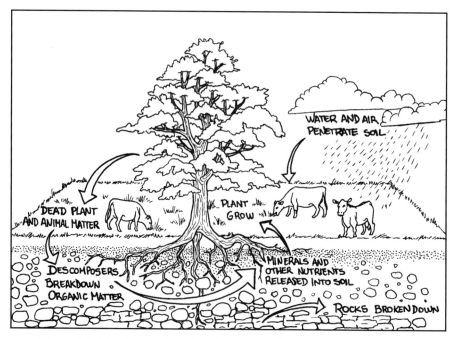

Figure 2.5 Nature recycles minerals, gases and water in every ecosystem.

All elements have their own nutrient cycles, but most are integrated as they involve the same processes and same organisms. The most important nutrient cycles that need to be considered in farming are the carbon, water, nitrogen and phosphorus cycles, although other elements that are essential for plant and animal development include potassium, calcium, magnesium and sulphur.

13. Enhance Ecological Succession

Ecological succession is a slow process where the plant (and animal) communities change over time. As an environment changes, there is a corresponding change in the numbers and types of species present. A classic example is when a block of land may be cleared for a house. If the house is never built, you would expect weeds (pioneers) to invade, then an occasional shrub appears and after a few years some large trees (climax species). Even the weeds may change from year to year. Bear in mind that weeds might have very good mycorrhizal associations and this is important to get them established or re-established to help repair the soil. Succession is driven by disturbance, whether that is caused by soil disruption or by fire, volcanic flows or any severe weather events that damage and change the environment.

When we plant a garden we incorporate the same strategies as nature. We put in fast growing pioneers (such as nitrogen-fixing cover crops and nursery shrubs) to improve and protect the soil, we plant a range of different sized plants to exploit sunlight or nutrient availability, put in our companion plants to help with pest control or attracting pollinators, and we end up with what we value as high return trees – the climax species such as oaks or nut and fruit trees.

Coupled with this, we use plant and time stacking, a multilayered planting regime (the seven or eight layers of a forest) and edge effects – all techniques to make a biodiverse and highly productive garden or food forest.

14. Integrate Livestock or Other Types of Animals in the System

Can you be a regenerative farmer and only grow plant crops? Yes. However, in almost every ecosystem in any part of the world animals are present. Some farmers, through choice or climate and landscape factors, are only croppers, some are dairy farmers, and some run sheep or cattle – with or without crops. Some farmers have realised that animals are hard work and are not worth the effort.

However, most prominent, well-known regenerative farmers or organisations see the value of stock as essential parts of the mix. The issue may be how you see and use animals in the farming operation, and this ranges from breeding, selling and slaughtering to simply using them as sources of weed control and to provide manure for the crops that follow.

What is essential though is good animal health and welfare, where animals are treated humanely, because animals should be allowed to have a dignified life.

15. Use Integrated Pest Management Strategies

Integrated pest management is using a range of strategies to control pests, disease and disorders. These strategies fall into four groups – chemical (sprinkling flour, a stomach poison, on brassicas to control white butterfly grubs), biological (using natural predators such as parasitic wasps to control pests), physical (picking off grubs from leaves) and cultural (clearing away fallen fruit, pruning dead wood).

Sometimes diseases and nutrient deficiencies can easily be corrected by using amendments and nutrient sprays (solutions of copper or lime sulphur). It might not be necessary to completely eliminate a pest or disease, but rather keep them to a minimum where their numbers have no real impact on the economical return of the produce.

Regenerative agriculture focuses on biological control methods, followed by cultural and physical controls, with chemical use the least favoured. Chemicals tend to be indiscriminate and threaten non-target organisms, so even the less harmful ones kill or deter beneficial insects as well.

16. Enhance Local Environments

Every local environment needs protection and many have severe degradation already. Much of the restoration works are carried out by local residents and community organisations on a volunteer basis, but often there is little funding and support from governments. It is crucial that we address the issues of poor water and air quality as well as deterioration of soils and landscapes.

In some areas of the world massive revegetation and rehabilitation pro-grammes must be rolled out to ensure the declining number of species have habitat, food and shelter. Planting local species in remnant bush and forest areas adjacent to farms enables greater biodiversity and interaction between all species, and provides ecosystem functions for the crops we grow and the animals we keep.

Figure 2.6 Before (L) and after (R) revegetation works.

Farmers see themselves as custodians of the land and rightly so, they have opportunities to mitigate carbon loss from our soils and reduce greenhouse gas emissions. This is becoming increasingly important when you consider that each person contributes about six tonnes of carbon dioxide each year from all the activities that humans engage in.

17. Work with and within the Community – Build Social Capital

Social capital is the relationships within a community where members work together to enable the society to function effectively. They might share similar values, engage in fellowship and endeavour to get along and trust each other. This social interaction shifts from competition to co-operation. So there is a shift in thinking from 'you or me' to 'you and me'.

Like a food web there are links between various members, bonds strengthen between family members and extended family members, and the community as a whole share a common culture. Successful regenerative enterprises can strengthen a community as more rural folk are engaged in supporting the farm, and the local regional community prospers.

18. Use Local and Renewable Resources, with Low Embodied Energy

People often say 'support local, buy local'. While that may not sometimes be the best value, it is imperative that we all acknowledge food miles and the great cost to transport goods across the globe. Fossil fuel availability is decreasing and progressively increasing in cost. Renewable energy that captures and converts the energy from the sun, wind, waves and fermentation of organic matter (biogas) is our future.

As we continue to mine and exhaust minerals from the earth, and where recycling of these is not always possible, we will come to rely on materials that have low embodied energy. Embodied energy is the energy used to make products, including the mining, processing, manufacturing and transportation from where they were first extracted to their final destination. For example, an aluminium window has high embodied energy whereas timber milled at the local sawmill has much lower embodied energy. That means that building with materials such as timber, rammed earth and straw bales will become more common.

We need to take more advantage of local knowledge and practices, and recognise the skills, knowledge and abilities that the members in our local community have and are so willing to share.

19. Work Towards Profitability

Every business needs to make a profit, even a modest one, or provide enough income to support family members. Some people view farming as strictly a business, which is a shame. It offers so much more, and farming is truly a right livelihood operation.

To ensure that farming is profitable and provides income all year round, farmers could diversify the income from different enterprises. If a farmer grows wheat, can they invest in a small grinding mill and offer boutique flour to the local community? If dams are present can they be stocked with crustaceans or fish? If there are large stands of native vegetation present can beehives be installed and managed, even by a local apiarist, to provide a pollination service for your crops as well as supplementary income from honey sales?

As the general buying public becomes more educated about food quality and food production there will be a greater demand for products that are chemical-free, hormone-free and produced from the ethical treatment of animals.

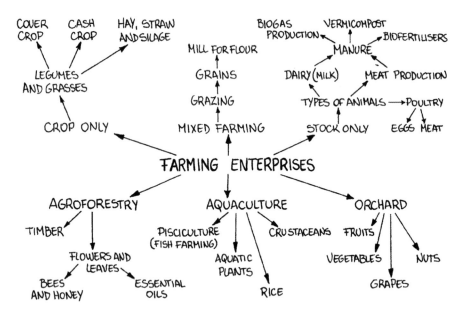

Figure 2.7 A range of different farming enterprises are possible.

20. Slow and Steady

Seldom do farmers have all the resources and finance to make a significant shift in their farming operations and expansion into new enterprises. Part of the design for resilience that was discussed earlier is the need to produce a five to ten year plan of implementation. Each year do a little more, but set some goals on priorities and work towards achieving these.

Slow and steady also allows the adoption of new technologies and new techniques as they arise. Your plans will change, but that's alright. Ecological restoration takes time and resources, but as they say, the longest journey starts with the first step.

3

All about Soil

All life on Earth depends on the soil. All human civilisations depend on the soil. The importance of good soil cannot be under-estimated and we should not see soil as a commodity, something we use and abuse with scant regard. We need to appreciate the natural capital of the living soil: We are technically made of topsoil – that's where all of our food that we eat is grown and what is used to make our bodies.

The perception of soil varies amongst community members. It may be seen as a nuisance by engineers who are building roads or a blessing by farmers and homeowners who want to grow plants. A farmer earns a living from the soil they tend and they pay more attention to it, so to them soil is indispensable.

However, every farm is degraded in some way, some less so, but many severely so. Degradation might be in the way of erosion, topsoil loss, low organic matter, poor soil structure and compaction. The problem we find is that building soil takes time. Nature might take 200 years or more to make one centimetre of soil. Even though soil is created faster in moist tropical areas, it is a lot slower in drier temperate climates. It is imperative that we speed this process up in all locations around the globe.

Soil Health

Healthy soils are the key to growing healthy food. Soil health is the status of the soil in terms of its composition, nutrient balance and soil life, all contributing to the functions that soils have. Soil health is about the ability of the soil to maintain biodiversity, productivity and ecosystem services in an environment. Optimum soil health is the interplay and continuous interaction between the biological, chemical and physical soil factors, and the ideal balance can be difficult to attain as the biological factors are so variable in different soils and different climates.

Soil is created by biological activity as well as chemical and physical processes. Biological activity acts on weathered soil particles to cause vertical separation. Here, soil biology can influence the soil profile, increasing organic matter and soil carbon in the lower regions (but also throughout the profile). We need to get the appropriate soil biology right for a particular plant species, as nature's ability to replenish exported and depleted minerals is through biological activity in the soil, so there is less need for supplementary fertilisers.

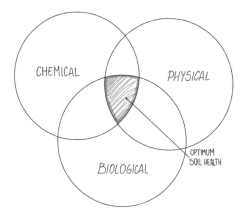

Figure 3.1 Soil health is a combination of
physical, chemical and biological soil properties.

The dynamic interaction between air, water, living organisms, geology (processes that produce soil grains), topography (slope and elevation), chemical
processes (that change mineral forms) and climate (temperature and rainfall) all
contribute to soil formation.

Soil health has to be made by the farmer and managed by the farmer to
ensure it changes in a positive direction. There are many approaches that help to
improve soil health, but we should rely on the six Ms of healthy soil: Moisture,
Microorganisms, Minerals (nutrients), Matter (organic), Mixing (air/oxygen)
and Make-up (soil structure and composition). Each of these are discussed in
more detail throughout this chapter, but not under those headings.

Many of the principles that were outlined in the previous chapter relate to
soil health. These include limiting disturbance (no till), keeping the soil covered
(with plant residue at least), building diversity (using various crop types such as
cool season grass, warm season broadleaf and so on), and keeping living roots
in the soil (having some roots in the soil keeps fungi and microbes alive). When
the soil is not disturbed through tillage practices, organic matter can build-up,
microbes flourish and soil health increases. The consequences of poor soil health
are shown in Figure 3.2.

You can often confirm a healthy soil by petrichor – the smell of rain. Most of
us can smell the earthy scent when rain falls on dry soil. This is a real phenomenon which occurs when raindrops hit the soil and various gases and aerosols
are released into the air. These substances are produced by soil bacteria and certain plants that excrete an oil during dry periods. Humans have an exceptional
sense of smell of these substances (no-one understands why), so if we move over
ground even after a light rain, we can detect them. Denuded soils do not possess
this anomaly and they simply indicate low soil microbiology.

Many farming agricultural landscapes are brittle and fragile. We probably
don't appreciate how quick soils decline, especially in brittle landscapes. Brittle-
ness is a measure of how fragile and dry a soil may be. Non-brittle soils have good

year-round moisture (through rain, snow or humidity) whereas brittle areas have a dry season when rainfall is low, irregular or non-existent for some months. Prolonged rest can heal non-brittle soils whereas prolonged rest in brittle soils leads to desertification. Most farming land is probably somewhere in-between non-brittle and brittle landscapes.

In a non-brittle landscape, such as a rainforest, there is enough water to maintain the soil microbes and to ensure good organic matter breakdown and nutrient cycling. Brittle environments have uneven distribution of soil moisture so you get dry plants, dead microbes, dormancy and slowed ecosystem processes.

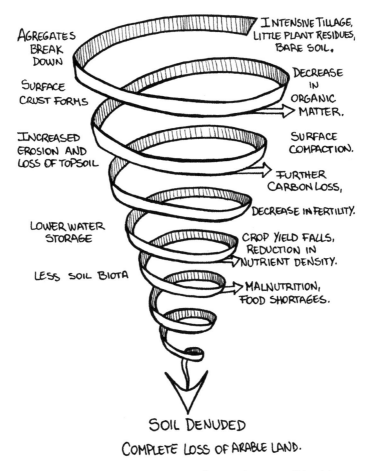

Figure 3.2 The downward spiral towards poor soil health.

Besides brittle landscapes, farmers have to deal with lots of other soil problems. Soil acidification is a widespread natural phenomenon, especially in regions with medium to high rainfall as water dissolves some of the minerals that create acidic conditions. Agricultural production systems can accelerate soil acidification processes through the removal of agricultural produce from the land, decreasing levels of organic matter and through addition of fertilisers and soil amendments that can either acidify soil or make it more alkaline. Natural soil acidification takes hundreds or thousands of years but agricultural practices acidify soils after only a few years.

Some of the acidity on farmland is due to nitrogen fertilisers. These can make soils quite acidic, especially using ammonium compounds (e.g. ammonium nitrate, ammonium sulphate) although using urea and other nitrate compounds (sodium and calcium) can increase the acidity (lower pH) too. Phosphate fertilisers generally have less impact on soil acidity, but certainly do contribute to changing soil conditions.

With the new wave of agriculture being promoted throughout the world post-1950s the thinking was that most soil-related problems, such as particular deficiencies, could be solved with external inputs, such as adding fertiliser. If the soil was dry add irrigation, if compacted use some machinery to break it up, and if there was a pest attack apply a pesticide. However, the ability to tell the difference between what the underlying problem is, and what is only a symptom of a problem is essential to deciding on the best course of action. Modern science has a clear message: Get the soil health right, including the amount of soil carbon and other nutrients as well as the soil life, then most of these types of problems disappear. There is also a strong link between soil health and human health, but industrial agriculture has broken that link.

Composition and Structure

Any soil has chemical, physical and biological properties, and the optimum soil fertility and structure is an intricate balance between organics, minerals and soil type and composition. Soil chemical properties include the cation exchange capacity, soil pH and the binding of soil minerals such as phosphorus to clay. Physical properties may include particle size, structure and stability of the soil, and water retention. Biological properties include the sources of organic matter (providing energy), and all of the myriad of organisms that live in the soil. Let's have a look at each of these in turn.

Physical soil properties include:

- **Texture** – that's the feel of a particular soil and it is a reflection of the size and varying (proportional) amounts of the three main grains or particles of sand, silt and clay. These simply vary in size, with sand the largest particle to about 2mm diameter, then lower than 0.05mm for silt and finally clay which is lower than 0.002mm in diameter. The composition of a soil is the particular combination of these three mineral particles and the naming of the soil textural class is illustrated in Figure 3.3. The soil triangle (right) can help you determine the composition of the soil. In this figure, an example is shown.

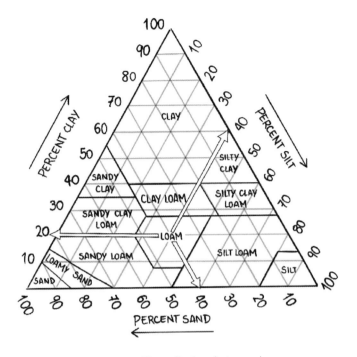

Figure 3.3 The soil triangle is used to determine the composition of particles in a soil.

You draw lines parallel to the sides of the triangle in the same direction as the arrows on the sides. Where your line cuts the sides of the triangle is the percentage of sand, clay or silt. With the example that is shown, loam is about 20% clay, 40% silt and 40% sand. You can use soil sieves to measure the amounts of clay, silt and sand and then determine what soil type you have.

Sandy soils feel very gritty, silty soils may be silky, while clay soils feel smooth or sticky. The feel of a soil sample helps to determine the type of soil it may be.

Some confusion may also arise when we use particular terms to describe soils. For example, we often talk about lighter soils and heavier soils. We are not talking so much about the weight of soil but rather its composition. We say sand is a lighter soil while clay is a heavier soil, but strangely, sand weighs more than clay for the same volume. The density of sand is higher than clay as there is more pore space in clay (even though the pore spaces are much smaller). The density of dry sand might be 1.5g/cm³ while clay is 1.2g/cm³ with loam somewhere in between at 1.35g/cm³. Using the numbers, if the soil was 1mm thick over 1 ha (10,000m²) this equals 10m³ (10,000 L) which is about 13-15 tonnes.

- **Structure** – the ability of those soil particles to form aggregates which hold onto the organic matter and nutrients. Soil is a spongy structure and good soil structure results in soil crumbs – little lumps of soil particles and organic matter glued together by microbes. These irregular-shaped lumps don't fit close together so air gaps between them allow oxygen to diffuse through the soil, providing oxygen for all soil life.

Besides this, healthy soil contains nematodes, earthworms, protozoa, termites, beetles and fungi among others which also build 'tunnels' (for want of a better term) which enable oxygen to move more freely too.

These aggregates, also called peds, often form distinctive shapes such as columnar or platy. Plain sand is single-grained and doesn't form an aggregate. Some aggregates are formed by the action of bacteria or fungi or when particles are attracted to each other through electrostatic charges or when other minerals or microbes glue the soil particles together into small lumps. When you hold a sample of soil and you find aggregates it might mean that the soil has good drainage or that it is reasonably stable, and both of these features are what we want in agricultural soils.

- **Porosity** – which is the pore spaces between the soil particles. This enables air and water to permeate into these spaces and throughout the soil, providing the plant roots and soil life with both of these essential requirements for life. A well-structured soil might have 20% of pore spaces in it.

  As mentioned above, the air spaces make the soil like a sponge and when it rains, water can penetrate and be stored there.

- **Water infiltration or permeability** – the movement of water through the soil. Permeability is particularly important if we want to hold as much rainfall in the soil as possible and allow it to move easily through the soil where it is needed.

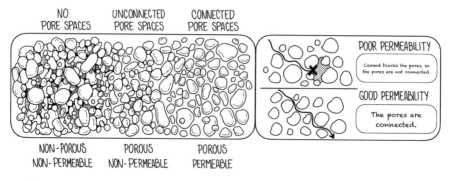

Figure 3.4 (L) Pore spaces in soil. (R) The movement of water through a soil.

- **Colour** – soil colour is caused by the minerals present in the soil. For example, iron causes yellow, orange or red-brown soils. Dark soils typically indicate high organic matter and this is what we want our agricultural soils to have.

· **Soil moisture** is simply the amount of water in a soil. It is important to get water stored in the soil and then made available to plants as required.

It is not only how much rain falls on the land, but how much infiltrates the soil and can be stored there. We need to condition the land to better use the rain that falls and not be too concerned about when it falls or how much falls on the day. Associated with rainfall events is run-off.

Different soils and soil covers have different run-off coefficients. If you imagine that a surface is concrete then the coefficient would be one as all rain runs away and none sinks into the ground. Bare soil might have a run-off coefficient of 0.7 while a grassed pasture cover might be 0.3. Compacted clay soil has a high run-off coefficient (0.8) so there is less infiltration, resulting in dry soils and low soil life.

Water infiltration also depends on the rainfall event intensity and duration. Light rains might sink into the soil, and rain over 120mm/hr or 2mm/minute may cause run-off, with only small amounts infiltrating the surface.

The dilemma for farmers is do they build drains to harvest run-off and store this in dams or are they best to devise ways to minimise run-off and get the water into the soil. While there is a need to collect water in dams to make this available to stock, our thinking should revolve around less run-off and more capture. We simply need to get more water into the soil when it rains and get the soil to hang on to it for as long as possible so that this can be available to plants.

Often we want to have some indication about the water holding capacity of the soil, and this idea is examined in Figure 3.5.

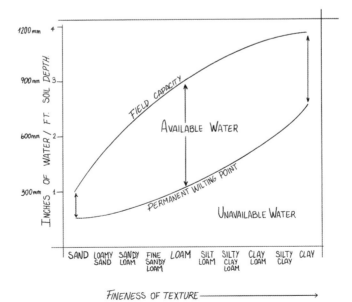

Figure 3.5 Available water in different types of soils.

This diagram shows how different soils hold and release different amounts of water to plants. The field capacity is how much water can be held in that particular soil. The wilting point is basically how much water is left in the soil that plants cannot access.

The difference between how much is stored and how much is withheld is the available water to plants (also known as Readily Available Water). Looking at Figure 3.5 you can see that loam holds a fair bit and withholds a little so the available water is high – higher than clay and sand.

Sandy soils hold less water than clayey soils and tend to dry out faster. On the other hand clayey soil tends to hold this water and not release it or make it available to plants. So we aim for loam that is a nice mix of clay, sand and silt because of potential water availability to plants.

Organic matter also affects water storage. A 1% increase in organic matter can result in anywhere from 1 to 5% increase in water-holding capacity. This equates to about 5 L/m^2 or about 50 tonnes/ha. Organic matter can store two to five times its own weight in water, and biological farmed soils hold more water too.

Not only is soil water necessary for plant and cell function it is also crucial for microbe movement through the soil, biological decomposition and nutrient release. How different would it be if we enabled soils to absorb more water and store it longer, rather than the disaster relief that governments provide for large scale flooding one year and then drought the next.

- **Compaction** – how much soil particles are pressed together, thus reducing the soil pore spaces. Heavy machinery, such as tractors, harvesters and trucks all compact the soil, but so does stock including sheep, cattle and horses.

  Compacted soils do not allow water and air to enter and may even become waterlogged on the surface. We tend to underestimate the importance of oxygen in the soil. All plants, animals and other organisms need oxygen for respiration. If a soil is compacted, has too little pore space, or is waterlogged (air replaced by water) then the soil becomes anaerobic, organisms die, plant roots suffer and nutrient uptake is seriously impaired. When air levels are low in a soil, it becomes anoxic and particular bacteria are mobilised to get the energy they need from changing soluble nitrates and nitrites into nitrogen gas, which is lost to the atmosphere. Soil nitrogen levels fall dramatically in compacted soil.

  Furthermore, soils that have too high a clay content tend to change from winter mush to summer concrete, and a compacted soil is the right condition for some weed growth. Nature always has solutions.

  Compacted soils need to be broken up, and this may conflict with the idea of not disturbing the soil surface. The use of special ploughs (also plows) that are dragged through the soil, breaking up the hardpan that is maybe 300 to 500 mm below the ground surface are commonly used. Examples of these chisel ploughs, as they were called, include the Yeoman's

(Keyline) plough, the Wallace plough, the Agrowplow and Kuhn chisel
plow. These ploughs drag tynes under the surface, endeavouring to break
up the hardpan, but with the least destruction to the soil surface.

Conventional ploughing, with a mouldboard plough or a disc plough,
turns the soil and releases stored carbon. When you plough you bring
organic material to the surface and then it oxidises, so everything you had
stored goes into the atmosphere.

Some soils will benefit from interim ripping and mechanical intervention to break-up any hardpans and surface crusting. As mentioned, the
use of chisel ploughs or similar can be used to kick-start the soil recovery
programme. However, as you plough you should spray some liquid fish
emulsion, seaweed kelp or composted weeds extract, biofertiliser, humate,
hydrolysate or similar substances as a way to increase the fungal content of
the soil. These solutions provide food for the soil critters, initially bacteria
and then other organisms that feed on them. You don't need to do this
every year, and most often it could be a few years at worst as you go deeper
and deeper to set up the soil to begin the change process.

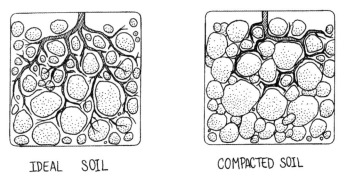

IDEAL   SOIL   COMPACTED SOIL

Figure 3.6 Compacted soil inhibits agricultural development.

Chemical properties may include:

- **pH** – the amount of acidity or alkalinity present in the soil. The pH scale
  is numbers 1 to 14, with 7 being neutral, neither acid nor base (alkaline).
  From one whole number to the next there is a tenfold difference in the
  concentration of hydrogen ions. For example, pH 5 is ten times more acidic
  than pH 6 and only one-tenth as acidic as pH 4. A pH of about 6.5 suits
  most plants and most nutrients are available too.

  Organisms tend to tolerate a pH range from about 4 to 9, so organisms
  may die or be seriously inhibited if levels more acidic or more alkaline
  than this are found in a soil. The pH of a soil can be changed by adding
  amendments. For example, adding limestone or lime makes a soil more

basic while adding peat or sulphur shifts the pH to a greater acidity level (lower pH).

Many farming soils are acidic. This can be due to natural causes or the addition of certain fertilisers. Superphosphate, for example, has a pH of about 1.5 – that is 100 times more acid than vinegar at pH 3.5. At this very low pH, iron, manganese and aluminium become more available in the soil and this causes toxicity. Superphosphate also has cadmium and fluoride so, if that is the case, maybe superphosphate should be used less.

- **Cation Exchange Capacity (CEC)** – the total number of cations or positive charges of all the nutrients that the soil, mainly clay and humus, could hold.

  Although clay particles are very small they form plates that are stacked on top of one another. When clay is surrounded by water the plates become negatively charged particles and a colloid is formed. A colloid is molecules that are dispersed throughout a solution. Sometimes these molecules and particles are large and when light is shone through the solution it is scattered and the particles can be seen.

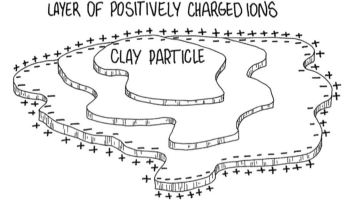

Figure 3.7 Clay forms negatively charged plates that can attract positive cations.

Clay and organic matter (humus) have negative charges on their surface and these can attract and hold onto positive charges (the cations) such as calcium, magnesium, sodium and potassium. These four are all alkaline ions (basic), but if the soil was slightly or more acidic then other cations such as hydrogen, iron, ammonium, manganese and aluminium are held in preference to the basic cations. In other words, these ions are exchangeable and can be held, replaced or released from the clay or humus. The humus found in organic-rich soils can hold about three times more cations than the best type of clay. Sand grains are not charged so can't hold minerals.

As clay particles are negative any negative anions (sulphate, nitrate) are not held and may be leached.

As plants take up particular nutrients, other cations take their place in the colloid. The stronger the colloid's negative charge, the greater its capacity to hold onto and exchange cations (hence the term Cation Exchange Capacity). The CEC is a good indicator of soil fertility, as a high CEC suggests many nutrients may be available to plants. A high CEC doesn't suggest or confirm that there are many nutrients in soils but rather the potential for whatever nutrients there are to be made available to plants.

CEC is measured in milliequivalents per 100 grams of soil (meq/100g). An 'meq' is the number of positive charges found in the soil. Each element has a particular charge (valency) and this is used to calculate the meq/100g soil from the actual concentration in parts per million, calculated by a laboratory test. For example, three common cations in a soil are calcium (Ca^2+), magnesium (Mg^2+) and potassium ($K+$). Calcium and magnesium have two charges (valency = 2) while potassium only has one. The gram equivalent weight is calculated by dividing an element's atomic weight by the valency. Calcium would have a value of 40÷2 = 20, while potassium would be 39÷1 = 39. A factor of ten is used to convert gram equivalent weight to milliequivalent/100g soil, then that figure is divided into the concentration from laboratory analysis. Sounds complicated, but the scientific calculation is relatively straightforward and typically yields values as shown in the table that below.

Table 3.1 Typical range of CEC values in soils

Soil texture	CEC range (meq/100g soil)
Sand	3-5
Loam	10-15
Clay	20-50
Organic matter (humus)	100-400

- **Salinity** is the amount of dissolved salt that is present in the soil. While we are concerned about sodium chloride (common table salt) in our water and on land, salinity includes all the other chemical substances as well. The use of fertilisers on our soils increases the salinity issue and may adversely affect plant growth and water quality. While nutrition is important for plant growth, some fertilisers are acidic in nature or might contain elements and compounds that are sparingly soluble and can be present in a soil for quite some time.

Some farming areas, typically close to seas or oceans, often have considerable salt levels in their soils. As an example, the wheatbelt area of

Western Australia is believed to have about 10,000 tonne/ha. Other places have rising salinity because of irrigation practices, or poor drainage combined with low rainfall and high evaporation. The Colorado River basin in western USA has this latter problem, where 5 million hectares (13 million acres) are affected.

- **Dispersion** is a property of soil, associated with clay, humus and cations. Soils that are easily dispersed respond well to adding gypsum. Dispersion occurs when soil particles, most often clay, move through the soil water. If the clay particles have sodium ions attached then they repel each other and move apart. We say that they have dispersed. If you want a dam to hold water then too much dispersion ability of the clay will make the dam leak. You need some dispersion to make the clay particles interlock and seal the dam, but too much sodium in the clay soil (a sodic clay) is detrimental.

    A simple soil test can be performed to ascertain the possible level of dispersion. Figures 3.8 and 3.9 show the set-ups that are easy to undertake.

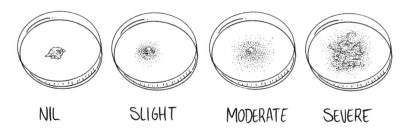

NIL   SLIGHT   MODERATE   SEVERE

Figure 3.8 Dispersion of a soil aggregate in water.

Once you carefully lower a soil aggregate into a distilled water solution in a petri dish or similar shallow jar, as shown in Figure 3.8, you observe any changes over the next day. If dispersion is likely to occur you should see some milkiness and murkiness within the first couple of hours. After 24 hours observe the extent of murkiness. The illustrations in Figure 3.8 provide some indication about the severity of dispersion.

Another variation to highlight dispersion is to shake a teaspoon of soil in a jar of water, as shown in Figure 3.9. Just shake for 20 seconds and then place the jar on a bench and do not move it for the next day. One jar should be distilled water and the other a gypsum solution.

Gypsum is Calcium Sulphate ($CaSO_4$), and to make a gypsum solution you add a tablespoon of gypsum to distilled water in a 500ml bottle and shake for half-a-minute. Allow this to settle and the gypsum to sink and the solution to become clear. Pour off the clear solution into another jar and use this for all tests. Gypsum is sparingly soluble so only a small amount dissolves at any one time (the solution is saturated). To make more

Figure 3.9 Dispersion and gypsum responsiveness. In each pair the left hand jar is soil in water, right hand jar is soil in gypsum solution. (L set) Soil 1. (R set) Soil 2.

gypsum solution just add more distilled water to the gypsum paste jar, shake, let settle and then pour off (decant) the clear solution. You will be able to make gypsum solution many times using that first tablespoon of gypsum powder.

Figure 3.9 shows two sets of jars. Soil 1 is dispersive because the soil in water is still murky after 24 hours. The right hand jar of soil 1 in the gypsum solution is clear. This means that the soil is gypsum responsive. In other words, if you want to reduce dispersion of a soil then adding gypsum to it will have a positive (beneficial) effect.

Soil 2 in Figure 3.9 shows no difference between the water and the gypsum solution. This suggests that the soil is already calcium based, so adding more calcium will have no effect. The soil is non-dispersive.

Typically you add gypsum at a rate of about one handful for every square metre (9 square feet). That works out to be about two tonne per hectare or 1800lbs per acre. However, it ideally needs to be applied to subsoil where it can be effective and not on top of the soil. Gypsum has a residual effect for about 10 years, so there is no need to keep adding it year after year. Just once and repeat, if you need to, 10 years later. The calcium in the gypsum makes the clay particles stick and clump together (flocculate) while a sodic clay causes dispersion, as shown in Figure 3.10.

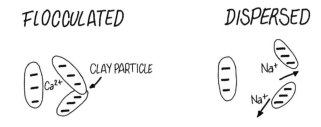

Figure 3.10 (L) Calcium makes clay particles clump together. (R) Sodium causes clay particles to spread apart and disperse.

You can also use your gypsum solution in a petri dish scenario. Make one dish water and the other gypsum solution. Add your soil aggregate to each and observe. The results will be similar to the jar tests.

- **Nutrients** – these are the substances that plants and all other soil life need for survival. Generally, nutrients are split into macronutrients and micronutrients. The ones that are important and essential are the macronutrients such as nitrogen, phosphorus, potassium, calcium, magnesium and sulphur.

  The micronutrients are those that are needed in small amounts such as iron, zinc, copper and molybdenum. Most elements are present in soil and rocks, but some are not readily available to plants. We need to recognise and acknowledge the importance of the biological aspects of soil, the soil biota, as these can change substances and supply these nutrients to plants.

  There are about two dozen elements that are fundamental for all life. The most common are carbon, hydrogen and oxygen, in various combinations. For example, hydrogen and oxygen make water, carbon and oxygen make carbon dioxide and carbon and hydrogen together make methane, which is a gas waste product. All of the other elements that include phosphorus, calcium, sodium and so on are essential to make all of the compounds that are found in plants and animals. These compounds include amino acids and proteins, fats and oils, sugars and all the other carbohydrates and substances that are utilised, such as enzymes and vitamins.

  While we do need to ensure plants have access to nutrients, a word of caution about artificial fertilisers. These tend to be soluble, so much is lost through leaching and this pollutes the groundwater and other waterways. Some farmers apply more than what is required simply to account for what is lost. This just adds extra input costs. As we will see shortly there are other mechanisms in soil that make nutrients available to plants. Typically, today's farming mindset is 'if you don't put in fertiliser nothing grows'. That is just not true and never has been.

  While soils usually have a full range of nutrients, with some limitations, these minerals are not always available to plants simply because the biology of the soil is not right. Biological properties are those related to the types of organisms present in the soil, and these could be microscopic bacteria right through to earthworms and larger animals. More on soil life is discussed later in this chapter (p.48). We don't often refer to carbon as a macronutrient, but the importance of carbon is discussed in the next section.

  Unfortunately, in the last few decades we have focussed on nitrogen, phosphorus and potassium (NPK) to improve plant yield without understanding the roles of all the other macro and micro nutrients and the interaction between plants and the community of soil organisms, all of which contribute to the plants structural integrity.

Soil Organic Matter (Carbon)

The non-mineral fraction of soil is referred to as organic matter, and although this is composed of many substances we normally express organic matter as organic carbon, as this is what we can easily measure. Topsoil organic carbon has generally much higher levels than that of the subsoil, but we need to get the subsoil holding more carbon, and encourage root growth and longer roots to allow more carbon at depth.

While carbon can be found in organic compounds (such as carbohydrates, amino acids and fatty acids) some does exist in inorganic compounds (such as charcoal and calcium carbonate). Soil organic carbon is the carbon-only fraction of the soil organic matter. Soil organic matter is a continuum of progressively decomposing organic compounds so we shouldn't think of it as this large amount of stable substances, but rather compounds in a state of flux. Soil organic matter is not just an assembly of large humic substances, but a full spectrum of substances that are undergoing continuous change. Farming needs to be more about managing carbon flows than managing carbon stocks.

Carbon is the key. When we think about carbon we visualise coal, charcoal or even the graphite in our pencils, a black solid lump which stays that way for a very long time. Solid carbon is basically insoluble in water so a lump of coal might take many thousands of years to decay and dissolve, while charcoal may take just tens of years. This is because charcoal is filled with lots of pores and microbes have a lot more surface area to work on.

The process of breaking down carbon by microbes is called mineralisation and results in the production of carbon dioxide. Often, microbes can act on carbon and change it into humus. The mineralisation of charcoal is sped up when a food source such as glucose is available. This food enables microbes to multiply and act on the charcoal.

The soil organic carbon doesn't have to be charcoal but all of the organic compounds present in the soil that the microbes and other organisms need as their source of energy. Soil contains more carbon than all of the earth's vegetation and atmosphere combined. If our soil continues to be degraded and depleted of minerals, microbes and moisture, and the temperature changes due to global warming, then we could expect large carbon losses from the soil to the atmosphere.

If plants can transfer more carbon into the soil than is lost from the soil then carbon can be stored in the soil. We call the process of storing carbon in the soil 'sequestration', which is one of the aims of regenerative agriculture. However, environmental disturbance may shift the equilibrium and cause more carbon to be lost and, unfortunately, contribute to a greater greenhouse gas problem. Carbon sequestration is also variable, affected by climate and weather, soil types, type of ground cover and the amount of soil biota – all changing the amount of carbon that is locked away.

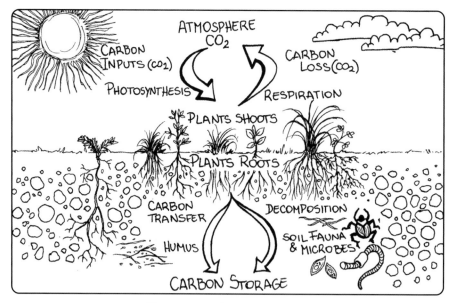

Figure 3.11 The carbon cycle.

Figure 3.11 shows the carbon cycle. Here you can see how carbon is transferred from air to soil via plants and then from the soil back into the air through various processes such as photosynthesis, respiration and decomposition. Some of the carbon is stored in the soil, some stored in plant tissue and the balance is found in other organisms and the atmosphere.

Carbon accumulation could be constrained when temperatures rise over 20°C and when rainfall is less than 400mm, and high temperatures and low rainfall often account for more variation in carbon stocks than land use and management practices.

Historical clearing of native vegetation and subsequent cultivation of semi-arid dryland cropping regions has resulted in substantial losses of organic carbon from soils. Semi-arid areas are suggested to be the most sensitive to global warming and most likely to reflect rapid climate change. With more intensive cropping, continued tillage, low retention of plant residues and increased soil erosion, the continual decline in soil organic carbon has occurred.

Carbon loss is affected by land use practices. When properties transition from pasture-dominated landscapes to grain cropping or when animal enterprises are being reduced, as stock prices and profitability falls, then soil carbon decreases.

Soil organic carbon is biologically active and is utilised by organisms as energy stores for the production of new compounds that contribute to growth and function. It is also important to stabilise some of the carbon below-ground and to enable the various compounds to move through the myriad of microorganisms.

We need as much carbon below-ground as possible, whereas most farmers aim to get as much carbon above-ground in their crops. The problem is that we tend to focus on what is going above-ground and ignore what is happening below-ground. Everyone has to grasp the need for balance – carbon is needed both below and above-ground when growing plants and managing residue, but still understanding and acknowledging that carbon (as carbon dioxide) is continually lost from the soil.

The carbon cycle should not be seen in isolation and separate from the nitrogen cycle, water cycle, phosphorus cycle and many other mineral cycles. They collectively form, not a food web, but a cycle web. Mineral cycles all interact. Although we draw a distinct cycle for a particular element, these really should be seen and understood as integrated and interdependent.

Fungi and microbes are involved in the transfer and storage of carbon in soils. Active beneficial fungi, the mycorrhizal fungi, permit more than 50% additional carbon to be stored in the soil, so their presence is crucial to soil health and to mitigate climate change.

As a general rule-of-thumb, for every tonne of carbon (C) in soil organic matter, about 100kg (15.7 stone) of nitrogen, 15kg (2.3 stone) of phosphorus and 15kg (2.3 stone) of sulphur become available to plants as the organic matter is broken down. So if you know the carbon content or at least the total organic matter content of a soil you can work out the potential supply of nutrients to plants. For example, there are about 2,000 tonnes of soil per hectare (890 tons per acre) in the top 150mm (6in) of soil. We can estimate the soil organic matter percentage by multiplying the organic carbon result by 1.72. This is a fair approximation, but assumes a 58% carbon content in organic matter.

If the organic carbon content is 2% (3.3% organic matter) then we could expect 30 tonnes of carbon (C) per hectare and, in turn, 3kg nitrogen (N), 0.5kg phosphorus (P) and 0.5kg sulphur (S) (73lbs C, 7.3lbs N, 0.6lbs P and 0.6lbs S per acre).

Since carbon dioxide is an important greenhouse gas and contributes to global warming and climate change, we need to reduce its concentration in the atmosphere. This can be achieved by simply doing two things: Plant more trees and improve our soils. All soil types respond to activities to increase soil carbon and again, carbon is the key to soil fertility, soil health and water holding capacity.

Basically, we need to get more carbon in our soils and keep some of this locked in so it can't escape. Anything that removes carbon from the atmosphere is called a 'sink', so to combat climate change we need a very large sink, and lock and store carbon in forest and grasslands and in the soil itself. Soil is both a sink and a source, but our efforts should be to sink and sequester carbon in the soil.

The three pathways to get carbon into the soil are decomposing litter (which is short-lived carbon), the sloughing of roots when the plant top is grazed or mowed and the liquid carbon pathway (which is the long-term carbon source promoted by Dr. Christine Jones). These sources run the soil sections of the various nutrient cycles. The liquid carbon pathway is probably not the best scientific way to describe the sugars and other carbon substances in the exudates that feed the soil and microbes. It is not 'liquid carbon', as you might visualise liquid water or molten steel, but dissolved carbon substances in solution.

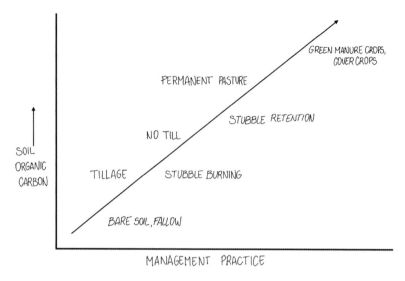

Figure 3.12 The effects of management practices on levels of soil organic carbon.

In any case, if soil carbon increases we can definitely say that the system is regenerative and conversely, if soil carbon decreases, then the system is degenerative.

Generally, since forests are tall and a lot more woody than grasslands, forests can store about ten to twenty times more carbon than grasslands. It has been estimated that up to two tonnes of carbon can be removed from the atmosphere every year and stored in each hectare of woody plants. At any one time about five tonnes of carbon are stored in grasslands and about 100 tonnes in forests in each hectare. It is easy to see that by planting more trees and stopping unnecessary clearing, then greater carbon sequestration is possible. Furthermore, by having greater plant cover and density, as well as improved soils with high levels of organic matter, the amount of carbon dioxide removed from the atmosphere can be significant.

Along with the carbon cycle, other elements such as nitrogen, phosphorus and potassium, as well as water, also have cycles. The nitrogen cycle, in particular, is important as all amino acids, proteins and all genes (made up of DNA and RNA) contain nitrogen. Many of the microorganisms in soil are involved in changing atmospheric nitrogen gas into a range of substances which ultimately plants take up and use.

Nitrogen is a nutrient essential to all life but often is deficient in soils used for crop and tree production. While nitrogen gas (N_2) is very common in the atmosphere (about 78% of air is nitrogen) it is only available to plants as either ammonium (NH_{4+}) or nitrate (NO_{3-}). This means that nitrogen gas has to be transformed into these other forms so plants can absorb and utilise them.

Nitrogen in the soil exists in many forms and these forms can change (transform) very easily from one to another with the help of microbes. Most soil nitrogen (90%) is bound in organic matter and present in soil organisms, but this is not directly available to plants, other than through the rhizophagy cycle that is discussed in the next chapter (p.65) on plant function.

Some of this organic matter can be broken down into simple amino acids or further changed into inorganic forms such as ammonium, nitrite (NO_2-) and nitrate. A number of chemical processes occur to change nitrogen gas into other forms. For example, nitrogen-fixing bacteria and lightning can change nitrogen gas into ammonium, then other bacteria convert ammonium to nitrite and then onto nitrate (nitrification) and finally, other bacteria convert some of these substances back again into nitrogen gas (denitrification) which is released to the atmosphere. All of these processes are part of the nitrogen cycle which is very complex, but it is an important nutrient cycle that all life depends on.

Nitrogen fixation is an anaerobic process. The presence of oxygen changes the mechanism and types of substances that are produced. Some nitrogen fixation can occur with symbiotic bacteria, and other organisms can take the exudates from plants and make nitrogen compounds. So we don't have a nitrogen limited plant, but a nitrogen availability (to plants) problem.

What reactions take place in the nitrogen cycle depends on several factors. For example, in acidic soils of pH less than 6, microbes immobilise (lock up or remove) ammonium while in alkaline soils of pH greater than 7, nitrate is immobilised before ammonium. Furthermore, ammonium is positively charged so this is easily held by the negative charges on organic matter and clay particles whereas nitrate is negatively charged, not held by soil and so can move below the root zones of plants and be lost to the system and wind up in underground water.

In waterlogged soils denitrification is rapid so nitrogen (as gas) is lost from the soil, and the ratio of carbon to nitrogen (C:N) determines whether carbon or nitrogen is lost. For example, if the C:N ratio is low then nitrogen is mineralised and this increases the plant-available nitrogen content. High C:N ratios will immobilise nitrogen and therefore reduce the amount available to plants. Our aim is not to do anything that will deplete the soil carbon, such as when nitrogen fertiliser is added it reduces the carbon content.

As plants require quite a lot of nitrogen, and the levels can change in different soil types and in different climates, then unless this is replaced or added in some way, continual cropping will eventually lead to an exponential decrease in soil nitrogen. This is compounded by the fact that nitrogen is generally found in low concentrations in the soil. For example, it may be only 1g/kg (0.1%) in sandy soils through to 2.5g/kg (0.25%) in clay soils.

Having healthy soils teaming with microbes, adding nitrogen-containing amendments, biofertilisers and other fertilisers, as well as planting nitrogen-fixing shrubs and trees all contribute to meet the nitrogen demand of plants. Similarly, without organic matter we would have wider ranges of acidity in the soil, increases in salinity and an imbalance of nutrients and, of course, soil erosion.

Organic matter is crucial to soil life, plant growth and food production. Even though soil organic matter is widely regarded as critical to soil function and plant productivity there is little recognition by farmers that climate and land management practices adversely affect carbon stocks.

Soil organic matter is mainly dead and decaying plants and animals (plant and animal residue, large size or particulate organic matter), but a small amount of organic matter is living tissue, such as roots and microbes. The level of organic matter in soils typically varies from 1 to 10%, although it may be as low as 0.3% in deserts and up to 15% in intensive dairy farms with permanent sod (soil cover) and water. A value of at least 5% is ideal to grow crops and plants. At this level the release of nitrogen and carbon from the organic matter is adequate without additional fertilisers being required. Most of the organic matter is located in the first 300mm (1ft) of soils.

Particulate carbon is fresh or decaying organisms and their wastes, is labile (easily broken down), and can change rapidly as decomposition occurs and particles are oxidised. Studies suggest that any negative changes in total soil carbon are associated with the rapidly transitioning particulate carbon fraction.

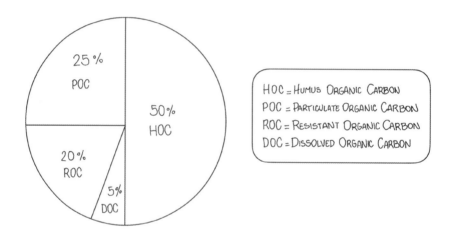

Figure 3.13 Fractions of organic carbon.

Another small fraction (about 5%) is dissolved organic matter, which is normally soluble root exudates (see later in this chapter, p.56). This has a high turnover as these substances are used all the time by the soil biota.

While organic matter plays an important role in agriculture, not all of it is available or useful in soil. About 10-20% of the organic matter in a soil is known as resistant organic matter as this contains substances that are difficult to break down (e.g. charcoal). The allocation of carbon to different fractions is strongly associated with the soil textural group (sand to clay) and pH. For example, the particulate organic carbon fraction is proportionally higher in clay than sand and total organic carbon decreases with increasing pH.

Much of the plant and animal organic matter in a soil is converted to humus by microorganisms (humic organic matter). Humus is a relatively chemically-stable substance made from a mix of complex organic compounds and minerals and held together with soil particles. It's like the soil grains are glued together by humus to make small lumps which we call peds or aggregates.

Humus is carbon-rich and predominantly exists as large organic substances, mainly carbon (60%) with smaller amounts of nitrogen, phosphorus and sulphur. As humus is broken down its components can be made into forms usable by plants, and thus humus is often called plant food. Humus is dark in colour due to its high carbon content.

Humus makes up about half to two-thirds of soil organic matter. The carbon-based complexes give soil structure, porosity, water holding capacity and cation and anion exchange. Humus can hold water seven times its own weight. However, the concept of humus being that stable dark material rich in carbon and nitrogen is flawed. This material is only generally observed when soil extraction with highly alkaline caustic soda (NaOH) produces a dark precipitate (humic acid), with the remaining soluble organic fraction called fulvic acid. The suite of hypothetical transformative processes have been collectively called humification.

Associated with humus is fulvic acid and humic acid, which are very active compounds of soil carbon. Fulvic acid is a complex, large organic molecule comprising carbon, hydrogen, oxygen, nitrogen and sulphur with multiple carboxyl groups (COOH) which release hydrogen ions (H+) and can form chelating complexes with metals such as iron, aluminium and copper. Fulvic acids have high exchange capacity. Fulvic acid is soluble in water at all pH levels and is believed to be a product from microbial metabolism, but seemingly not used as a carbon source by microbes.

Humic acid is a mix of organic compounds, such as sugars and proteins, comprising the same elements, but with a lower ratio of carbon to hydrogen than fulvic acid, so there are many aromatic rings (six carbon rings) and less carboxyl groups (less acidic). It is mainly insoluble in water, soluble in alkaline conditions (but not in acidic conditions), soluble in aromatic compounds (pesticides etc.) and is able to adhere to these substances in nature. Humic acid also forms complexes with heavy metals such as cadmium and lead (as chelating systems). Humic substances should be thought of as an alkali extract rather than suggesting a particular, distinct category of organic matter.

Alkali extracts of soil organic matter can be misleading as not all organic substances are extracted and the alkaline conditions ensure pigments and other similar substances are removed to become that dark humic material. It might be more beneficial if we extract water-soluble organic substances to better manage soil health rather than some mystical 'humus'.

Soil Life

Soils are alive. About one-quarter of all life exists in the soil. The workforce below our feet is very efficient, but little understood and appreciated. Each handful contains literally billions of bacteria. Soil is known to house bacteria, fungi, arthropods (insects, millipedes, centipedes and land crustaceans), nematodes, molluscs, burrowing earthworms, mammals and marsupials, and so many more thousands of plant and animal species.

Scientists think there are about ten million species of organisms on earth, but as we discover new species every day and we cannot, or have not, explored all of the ocean depths and every square metre on land, the number could be far higher. Either way, so far, about 75% of all species are found on land and 25% in the oceans and waterways. In terms of weight, plants contribute the most to total biomass than bacteria, fungi, archaea, protists, and finally all of the animals combined. We will examine bacteria and fungi shortly, but by way of explanation, archaea are single-celled organisms, much like bacteria, but with distinct features that set them apart. They live in low oxygen environments (anaerobic conditions), such as deep in soil, in water, in hot springs, in marshland and in the guts of animals. They have roles in fermentation and the production of methane, but some are also known to fix carbon, fix nitrogen and use sunlight as a source of energy. Archaea species are very common in soil.

Protists are single-celled organisms, although some form colonies or chains of these cells. Well-known common examples are amoeba, paramecium, euglena, slime moulds, diatoms and other simple types of algae. They are also known to cause various human diseases such as malaria, sleeping sickness and amoebic dysentery.

Protists are responsible for a large amount of photosynthesis that occurs on earth every day (algae), as decomposers and in recycling nutrients, and as important participants in aquatic and terrestrial food webs (both as food for others and eating others, such as bacteria). Protists are also very common in soil, and they can be seen in Figure 3.15 which shows an example of a soil food web.

Two of the most important groups of organisms in soil are the bacteria and fungi. Bacteria and fungi play essential roles in the breakdown of organic matter to create humus and other substances in the soil. Bacteria are single-celled organisms that are not true animals or plants, but are their own kingdom. They have thick cell walls like plants, but cannot make their own food and they move like animals. Bacteria are also unique as they do not have membranes around their nucleus and other internal parts. This feature labels them as prokaryotes (as are archaea), as opposed to all organisms called eukaryotes that do possess membrane-bound organelles (all protists, plants, animals and fungi).

Soil bacteria do many different types of jobs. Some break down organic matter into a range of substances that is the glue to make soil clumps and improve structure (allowing better movement of air and water throughout the soil), some are nitrogen-fixing and increase soil fertility, others release chemicals to inhibit nearby pathogens, some take the sugars and other substances that plants make and pass these on to other soil organisms including beneficial fungi, while others work on rocks to release essential nutrients which ultimately plants need.

Fungi are also in their own group of organisms – another kingdom. Again, they seem to be a blend of both plant and animal characteristics. Fungi have cell walls like plants, but the walls are made of chitin, a tough substance that is found in animals. Chitin is found in fish scales and the shells of crustaceans and molluscs. Fungi do not have internal vessels to transport water and nutrients throughout their structures and they cannot make their own food, so they must obtain their nutrition, food and energy by eating other things, and are similar to animals in this regard.

Fungi can be beneficial and also harmful. We eat mushrooms but toadstools are poisonous. There are three broad types of fungi – symbiotic (mutualistic), saprophytic and parasitic.

Mycorrhizal fungi are examples of symbiotic fungi, while yeasts and mush-rooms are saprophytes as they feed on dead or decaying matter. The mycorrhizal fungi are unique and their role in soil health is discussed in the next section.

There are some soil fungi that are not that helpful to plants. They are the patho-genic (disease-causing) or parasitic fungi which rob from plants, often causing their deaths. These include some moulds, smuts, soots, root rots and rusts which can seriously affect our food producing crops, such as cereals, vegetables and fruits.

There is a balance of bacteria and fungi in soils, but in many instances this shifts towards either a bacteria-dominated or a fungi-dominated soil. Propor-tionally more bacteria are found in agricultural and pastoral soils where grass, grazing animals and manure are common. They prefer lots of green manure (soft-leaved vegetation) and nitrogen-rich soils, so bacteria dominate in our vegetable gardens. However, bacteria dominated soils have low nutrient availability as the nutrients are locked up and immobilised. The soil can also be fine, structureless and prone to erosion.

Plant farms that have been highly cultivated or chemically farmed have bacte-ria dominated soils. Both soil disturbance and soil compaction shifts to bacterial dominance. When this occurs, opportunistic weeds invade. Weeds are the ulti-mate survivors. They have mechanisms that enable them to endure, survive and thrive, and allow them to disperse into new areas. We need to use our growing understanding of biology to farm better.

Fungi, on the other hand, dominate high carbon soils such as forests and where trees are found. Fungi don't like disturbance (soil turnover) so prefer mulched, woody forest floors. You know you have a fungal soil if you see mushrooms (all types) popping up after rain. Fungi prefer less cultivation so in a 'sleepy' soil mode there is less disturbance and this is more suited to fungi. To increase fungi you need to get a good fungal attack on the leaf litter and organic matter in the soil.

To work out whether a soil is bacteria or fungi dominated, we can measure the weight of all the fungi and all the bacteria (their biomass) and compare the two.

Agricultural soils might have a fungi to bacteria biomass ratio of 1:1, but forests could have a ratio tens to thousands times more (much more fungi compared to bacteria). These ratios translate to carbon storage and loss. For example, if the fungi:bacteria ratio is high then more carbon is stored and less carbon dioxide is released. Figure 3.14 provides some examples of typical fungi:bacteria ratios in various soils.

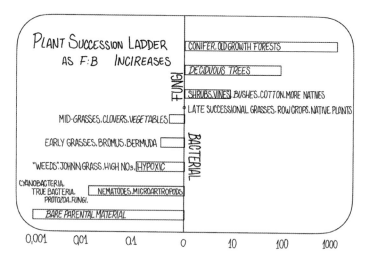

Figure 3.14 Examples of fungi to bacteria ratios.

For orchards, the fungi:bacteria ratio might be 10:1, grasses 1:1 or 2:1, while woody forests 100:1. In tilled soil it could be 1:100. Ideally, in a regenerative agriculture system the ratio of fungi to bacteria is closer to 2:1 and up to 5:1.

If we practise no till in agricultural systems and the soil is left undisturbed then the fungi to bacteria ratio increases and we often find that more organisms prosper.

We often talk about a symbiosis between two organisms like algae and fungi in lichen or a plant and bacteria that provide nitrogen-fixation. However, when we look at the bigger picture, the plant and bacteria and other microbes are all involved in soil. So there is a symbiosis between all three, with soil providing porosity and air, water-holding capacity, and the storage and release of nutrients. The plants themselves exchange these things with the soil and the microbes, whether they are nitrogen-fixing or doing other important ecosystem functions, and all of these are integrated and working together to improve soil. The synergies and interactions between organisms are complex and little understood.

While fungi and bacteria share the role of decomposing organic matter, each group is affected by soil conditions such as pH and temperature, and each group has their own nutrient demand, turnover rate and interaction with other species. Soil pH influences carbon and nutrient availability as well as the solubility of metal ions, and it also affects the relative proportion of fungi and bacteria.

At pH of 4, microbes are not active, but as the pH increases (towards alkaline conditions) so does the organic carbon and soil nitrogen content. At pH 4.5 or so fungi are dominant but at pH 8 or so bacteria are dominant. There is an inverse relationship between these populations. As bacteria increase, fungi decrease and vice versa.

Bacterial growth is greatest when the pH is above 7 and fungal growth peaks at about 5, but less than 7 at least. However, both groups are severely affected by pH of 4 or less (very acidic) or pH of 9 or more (very alkaline or basic). For example, at very low pH (high acidity) levels much more aluminium is available and this may be toxic to microbes. So if we add lime or ash to soils to increase the pH the microbial populations change. It may mean that high hydrogen concentrations favour fungal growth while low hydrogen ion concentrations favour bacteria.

Scientists can also measure soil life by capturing the amount of carbon dioxide emitted from a soil. Higher carbon dioxide amounts indicate higher numbers of organisms. The carbon dioxide is given off when all of these organisms respire. Of course, scientists can also literally count the organisms by looking through a microscope. These techniques help paint a picture of soil life and soil health. You can ascertain how much soil life you have by sending off soil samples for a laboratory to perform the Haney test or similar tests which measure microbe activity and carbon dioxide generation from microbe respiration.

Larger animals also exist in the soil. Too often we are told about the baddies in the soil, like the nematodes that distort roots and hinder plant growth, and so much of our scientific focus is about how we can successfully control them. What people are not told about is all the useful nematode species, that break down organic matter, are integral parts of the soil food web and build soil structure. Many more insects, crustaceans and other arthropods, worms, reptiles and small mammals make the soil their home.

A special mention must be made about earthworms. Although earthworms are key players in the process of vermicomposting, microorganisms living in the earthworm's gut perform the actual decomposition of organic matter. As earthworms burrow through the soil they take in sand grains, which are used as grinding stones to break down the organic matter they ingest. Bacteria are then better able to act on the organic matter fragments and change them into much simpler substances.

Vermicompost is a nutrient-rich, dark material which has high water-holding capacity, high nutrient storage and a low C:N ratio. It is often used as an amendment for improving soil or to aid plant growth, but it also stimulates seed germination and development, flowering and fruit production in a variety of plant species. Vermicompost can also be placed in water to make a solution containing dissolved substances and this can be sprayed over the ground.

– play an essential role in carbon turnover as carbon is reduced in the digestion process of the organic matter, and is released as carbon dioxide. This, in turn, causes a significant reduction in the C:N ratio so proportionally more nitrogen is available.

Finally, a word of caution. Herbicides, pesticides and fungicides all significantly reduce nearly all life forms in the soil, and herbicides push the soil biology towards bacterial dominance, so fungi really suffer.

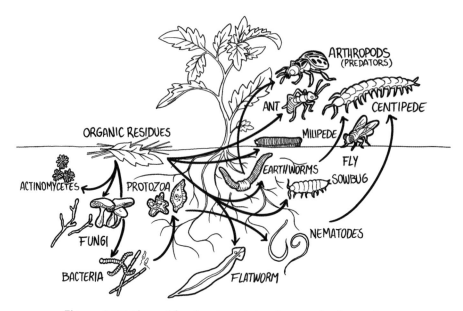

Figure 3.15 The soil food web contains thousands of organisms.

Good soil biology is what we need. It naturally ameliorates the pH and changes acidity towards neutral or slightly alkaline and all of the processes and interactions in the soil depend on the biodiversity that lives there.

Mycorrhiza Fungi

Fungi are often associated with disease: Think of tinea, ringworm and thrush (candida) in humans and black spot, mildew and rust in plants. However, one group of fungi are crucial to soil health. These are the mycorrhizal fungi, so named because they have feeding threads (hyphae) that attach to a plant through the region around the plant's roots (the rhizosphere). The rhizosphere is a thin, oxygen-rich layer around plant roots that itself is a unique and complex ecosystem for all of the microbes and other organisms that live there and interact with each other and the plant. Mycorrhizal fungi do not cause disease and they account for about 10% of all identified fungal species, although this may change as we discover and name more fungi and more mycorrhiza.

There are seven different types of mycorrhiza which can be split into two main groups: Those that have feeding threads inside the plant cells (endomycorrhiza) and those that do not penetrate the plant tissue but attach themselves to outside root surfaces (ectomycorrhiza). Ectomycorrhiza is associated with about 10% of plants, mainly trees and woody plants such as alders, eucalypts, pines, oaks, birch and conifers.

Another 10% of plants do not have a relationship with mycorrhizal fungi. These include brassicas (cabbage and canola), some common weeds, some nitrogen-fixers such as lupins, some members of the Proteaceae and reeds that live in water or moisture-laden soil. At times there may be rudimentary connections between these plants and mycorrhizal fungi but it varies. Often mycorrhizal fungi are not found in association with plants that live in waterlogged soil, or very dry or saline soils, disturbed sites (mining operations) or where soil fertility may be very high already. Other plants may be parasitic or carnivorous and do not need fungi helping them to survive.

While most of the seven types have symbiotic associations with plants, it is the Arbuscular mycorrhiza fungi (AMF) that we wish to focus on. Arbuscular are endomycorrhizal as they have special balloon-shaped vesicles and branching, tree-like threads called arbuscules (tree-shaped structures) through which materials are absorbed and exchanged. Because mycorrhizal fungi are inside plant cells, the exchange of substances is very efficient – little is lost to the environment and all substances are available to either organism.

AMF help plants by providing water and nutrients, such as phosphorus, calcium and sulphur, which they get from the soil. Plants, in turn, supply sugars, proteins, amino acids and other carbon substances that the fungi use for their own growth and for energy. This symbiosis is a mutualistic relationship because both benefit from the association. It's like a barter system where the mycorrhizal fungi give nutrients in exchange for carbon compounds from the plants. It's a win-win for both types of organisms.

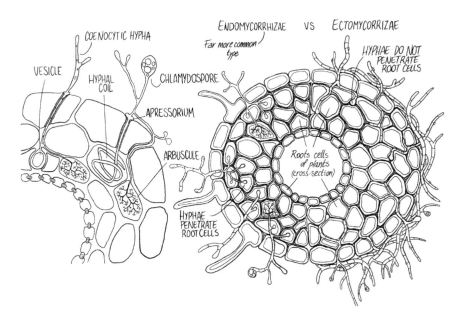

Figure 3.16 Arbuscular mycorrhiza fungi (AMF) are commonly found in association with many plants.

There are other benefits of these types of associations, including induced resistance to disease and pests. Induced resistance, when mycorrhizal fungi penetrates the plant root cell well, results in a plant response to fortify some plant roots which makes it difficult for pathogens to enter. The act of mycorrhizal fungi seemingly attacking the plant, ultimately causes the plant to improve its defence mechanisms.

As part of these symbiotic associations, plants exchange substances that the fungi need. It has been estimated that plants may allocate up to 20% of the photosynthetic substances they make to be passed out as exudates and secretions to their fungi partners and to other organisms in the surrounding rhizosphere.

The reason why this relationship is so important is that about 80% of all vascular plants (flowering plants and conifers), and possibly about 90% of all cultivated species, co-exist with arbuscular mycorrhiza, and this includes most fruit, vegetables, cereal crops and trees.

Mycorrhiza fungi have extensive hyphae. These very fine threads permeate the soil much further than plant roots and permit a plant to have essentially forty or fifty times more surface area to absorb water and nutrients. These minute threads are able to penetrate small soil pore spaces which plant roots cannot exploit.

Without mycorrhizal fungi, when we apply soluble fertilisers only a small fraction is available to plant roots with the majority being lost. There really is no guarantee that any applied chemicals will be actively absorbed by plant roots. The efficiency of uptake is very low so we have to keep adding more fertiliser to get the desired results.

A good example of issues that arise with only relying on artificial fertilisers is phosphorus. Phosphorus is easily locked up and adsorbed (especially by clay), and while it may be present in the soil it is just not available to plants. Having arbuscular mycorrhizal fungi present solves this problem because it can source phosphorus from the soil and provide it to plants. Phosphorus metabolic bacteria are also known to attach to mycorrhizal fungi, so this adds another layer to the intricate relationships between soil organisms.

We just don't know enough about the biochemical signals between organisms to fully appreciate the complexity of nature, nor how we can use this knowledge to make growing plants and getting high yields so much easier.

The presence of mycorrhiza fungi also provides other benefits to plants and to the soil. They enable the plants to thrive in adverse conditions such as drought and salt stress, they mobilise forms of nutrients that are not directly available to roots, they increase the host plant's rate of gaseous exchange and photosynthesis (and therefore growth). Yields of cereal crops are much higher when associated with AMF, they improve water movement through the soil, and they improve a plant's resistance to disease. Fungal hyphae attach to root hairs and help stabilise the soil aggregates. This enables better water infiltration as well as improved interaction between all soil organisms.

Some of the mechanisms of how this all works is unclear but researchers have found the mycorrhizal fungi produce a range of substances, with the main one being glomalin which is a sugar protein. Glomalin helps to hold soil particles together, enabling better soil stabilisation and water holding capacity. Glomalin has been described as the soil's superglue. It enables clay, organic matter, sand and silt to stick together as aggregates, small clumps that enable water penetration into the soil as well as making carbon available to microorganisms and plants. Glomalin also covers the aggregate surface to prevent the rapid movement of water into the clump which can cause the aggregate to fall apart as air is released. We need the aggregates to be whole, so plant roots can access nutrients via the mycorrhizal fungi which link the aggregates to the plant. Glomalin is also a store for carbon, protects the fungi hyphae and has been shown to bind with heavy metals, which may contribute to the possible bioremediation of contaminated sites.

Mycorrhiza are fragile, and they do not grow in a laboratory – they only survive when attached to healthy plants. While a more diverse range of mycorrhiza are found in natural plant communities than in any monoculture agricultural system, the hyphae of the mycorrhiza fungi are easily damaged when we cultivate or disturb the soil, spray chemicals to kill weeds, or apply soil fumigants.

Furthermore, if superphosphate or other similar manufactured phosphate fertiliser is used excessively on the soil, acidification can occur (superphosphate is acidic in nature) and mycorrhizal fungi may disappear completely. Simply adding lime to correct the pH may not allow the fungi to return and, if applied indiscriminately, can make the soil too alkaline.

Unfortunately, it is not as simple as just adding organic matter or various amendments to denuded soils if there are no mycorrhizal fungi present. Plants, animals and microorganisms (including fungi) all evolved together in intricate relationships, so you need to ensure that these relationships thrive if you want a regenerative system. Modern agriculture often decouples those relationships and it is our job to make them active again.

Artificial fertilisers don't seem to be detrimental to mycorrhiza fungi, unless they are very acidic or contain toxic elements. High concentrations of some fertilisers do alter the composition of both bacteria and fungi species, including mycorrhizal populations, but taxon richness seems to be maintained. On the other hand, fungicides are detrimental to mycorrhizal fungi and soil health.

However, we shouldn't add lots of chemical fertilisers when we want mycorrhizal fungi to proliferate. If nutrients are readily available and plants do not make those connections with beneficial fungi, the plant takes what it wants from wherever it can. The beneficial fungi are crucial to slowly releasing nutrients to plants over their lifecycle, which is a lot better than receiving one large lot of soluble nutrients when the plants are young.

Finally, as commercial mycorrhizal fungi have been shown to not be always viable, it is prudent to ask for confirmation of viability testing before you buy. Spores can live for years but often become weak if too old. What is clear though, is if fungi are present in the soil then nutrient availability to plants increases.

Plant Root Exudates

Earlier we mentioned that plants release up to 20% of the substances they make through their roots to mycorrhizal fungi, the rhizosphere and to the microbes that exist in the surrounding soil. These substances are called exudates. While some exudates are food and carbon sources for these organisms, other exudates play different roles.

Root exudates are a complex mix of soluble organic substances that include amino acids, fatty acids and lipids, organic acids, sugars and other carbohydrates, growth factors and enzymes. Some of these exudates are released as responses to disease, mineral deficiency, physical damage to plant roots and as signals to other plants and organisms.

Even so, most exudates are either organic acids, sugars or amino acids. The organic acids, such as malic, acetic and citric acids, bind to metal ions and make these available to plants, as well as dissolving minerals (weathering) to break down rocks. More acids, for example, are released if there is a particular mineral deficiency and they have the ability to induce or cause solubilisation of some minerals, such as phosphorus.

Exudates are often seen as part of the liquid carbon pathway or the carbon gift. The sugars and amino acids in exudates are used by the soil microbes as food sources, although some amino acids help with increasing the mobility of micronutrients so these are also more readily available to plants.

Organic acids vary in size and composition. Some are small molecules with low molecular weight, partially soluble in water, and include citric, oxalic, malic and acetic acids. Larger molecules with higher molecular weights that vary in solubility include humic and fulvic acids.

Some exudates act as signals to microbes. These signals could be to attract rhizobia or mycorrhizal fungi, trigger spore germination, suppress pathogenic fungi or allow a parasitic flowering plant to grow and find its host plant. For example, some flavonoids may suppress pathogenic fungi while other flavonoids (e.g those released by legume roots) attract rhizobia.

Small amounts of phytohormones are excreted from plant roots. Phytohormones are chemical messengers that coordinate the cellular activities of plants. Many are plant growth regulators, and examples include jasmonic acid, salicylic acid, abscisic acid, ethylene, auxins and gibberellic acid. They have roles as growth-regulating and stress-related hormones, and are crucial for seed germination, growth of shoots, ripening of fruit and so on.

Root exudates can also have weed-suppressive allelopathy. Allelopathy is a term that describes the influence of a plant that gives off chemicals to attract or repel other organisms or to enhance germination and growth, or to limit these. Many crop species exude substances to limit weed growth, with brassicas a well-known example of being used for biofumigation efficacy. Many plants store these allelochemicals in their leaves and when the leaves fall and decompose the chemicals are released which affect nearby plants.

It is unclear what mechanisms plants use to drive and control exudate production and release, but simple diffusion from plant tissue into the surrounding soil is the most likely way they are released from root tips. As we find out more about plant exudation we may be able to further improve agricultural production and find better ways to store more carbon and reduce soil carbon dioxide emissions. The role of plant exudates and their effect on microbes is discussed further when the rhizophagy cycle, and the impact of the oxidation-reduction potential (redox) on soil, plants and microbes, is examined in the plant function chapter, p.82.

Soil Amendments

The terms ameliorant, amendment and soil conditioner are used to basically describe the same thing: any substance that improves the physical or chemical properties of a soil to enhance the growing conditions for plants. Amendment = ameliorant = conditioner.

The term amelioration, however, is used to describe a process: That of improving degraded or poor soils and making them more usable for agriculture. The aim is to change the growing conditions for plant roots and, in some cases, make more nutrients available. So let's use amendments to mean a substance, and amelioration as the process when soils change.

While soil amendments primarily improve the soil physicochemical properties, the addition of organic amendments will also facilitate increased microbial activity and diversity, and this is the secret to much healthier soils and plants.

A large percentage of agricultural soils have subsoil constraints that limit crop growth and agricultural productivity. These constraints include poorly structured subsoils that result from high clay content and compaction, high exchangeable sodium ($Na+$) concentrations (resulting in salinity issues), and impeded drainage and permeability (resulting in waterlogging and erosion). Soil constraints may be near the surface or in the subsoil and they might even be variable across a paddock or a property. These constraints may impede water movement into and through the soil, water storage in the soil or reduce water and nutrient availability to plants (reducing readily available water to a plant).

Soil amendments have been consistently shown to be beneficial. They can act on clayey soils to improve aggregation, porosity, aeration, drainage and root penetration. On sandy soils they can increase the water and nutrient holding capacity and storage. However, you need to use amendments with caution. For example, some composts and manures are high in salt so this is not helpful. Salt burn of plant roots is common from an over-application of salty soil amendments. Plant-based composts are generally low in salt and may be more appropriate in some instances.

To be effective, amendments must be thoroughly mixed into the soil. Ideally, the amendment should be hoed in well below the surface. While disturbing the soil and upsetting the soil biota, the application may be a one-off. Like many things, what we do is often a trade-off or compromise. The only exception is mulching. A mulch does cover the surface and is not worked into the soil. Its job is to inhibit weed growth, cool the soil and retain soil moisture.

Not all on-farm practices to improve soils are reliable or financially reward-ing. For example, gypsum is widely used as a soil amendment to change sodic soils and to improve drainage in clay-based soils. However, as discussed earlier, it is sparingly soluble (it dissolves a little bit at a time), so its action is slow and changes to soil are not immediate. In many cases, gypsum is applied to the soil surface so it may never reach the subsoil and its effect and the cost of the gypsum application is futile. Somehow we need to get the gypsum to the compacted sub-soil layer or at least in the soil rather than on top of the soil. In the same way, the incorporation of organic amendments deep within the subsoil will make signifi-cant improvements in crop growth and yield.

Some strategies may also be ineffective and cause more problems when under-taken. For example, it is not a simple matter to add sand to clay, thinking that this is the way to make loam. Adding sand to a clayey soil just creates a soil structure akin to concrete.

When choosing an amendment consider:

- How long the amendment will last in the soil and if it is slow or fast acting;

- The current soil texture and structure, and if that needs changing;

- Any relevant plant and crop sensitivities to salts (check salt content of amendment); and

- The pH of the amendment, as this may either ameliorate or worsen the soil condition.

Broadly, there are two categories of soil amendments: Chemical-based and biological-based.

Chemical-based Amendments

Chemical-based amendments are natural rock minerals, chemical salts or man-ufactured substances readily bought or obtained, and this includes fertilisers. Some of these chemicals are mined and available in their raw state, others are manufactured as by-products of industry or from other ingredients. Limestone, dolomite and rock phosphate are all natural substances which are mined and processed in some way, so while these substances can be used in organic farming, the extraction and use of these still has sustainability concerns.

Table 3.2 lists some chemical-based amendments and how they can be used.

Table 3.2 Chemical or mineral-based soil amendments

Common name	Chemical name	Uses
Gypsum	Calcium sulphate	Flocculates clay, provides calcium and sulphur as nutrients.
Bentonite (clay)	Hydrated sodium calcium aluminium silicate.	Increases the soil's ability to hold water, predominantly in sandy soils. This clay also has a property of polar attraction, so will draw water towards it, which reduces hydrophobic conditions.
Sulphur	Pure element (symbol S), yellow crystalline solid that is insoluble in water.	Dusting sulphur is a fine powder fungicide. Powdered sulphur is the coarser grade used to lower pH in soil gradually (makes more acidic).
Rock dust	Ground rock (e.g. granite, basalt) from quarries. Can be a fine dust or coarser-grained for ease of use. Contains a wide range of elements, both macro and micronutrients.	Replaces minerals in leached, poor quality or overworked soils. Very slow acting, long lasting.
Dolomite	Calcium and magnesium carbonates	Sparingly soluble. Increases soil pH (counteracts acidity). Source of calcium and magnesium for plants.
Limestone and Lime	Calcium carbonate Calcium oxide	Raises pH, source of calcium. Limestone is more gentle than lime.
Zeolite	Forms of crystalline aluminosilicate. Either formed from volcanic ash or can be artificially produced.	Can hold and exchange nutrients required by plants, making nutrients readily available.
Potash	General name for potassium-based salts including sulphate, nitrate, carbonate and chloride.	Adds potassium. All of the potassium salts can be used as fertiliser.
Rock phosphate	Calcium phosphate.	Naturally mined mineral that supplies phosphorus, a major plant nutrient.
Vermiculite	Made by heating mica. Aluminium iron magnesium silicates.	Lightweight, holds water, bulks soil.
Perlite	White volcanic expanded glass.	Inert substance, allows better aeration of soil. Porous, so absorbs water.

Biological-based Amendments

Biological-based amendments are those that are produced from once-living things. These substances can be made from dead and decaying organisms, burning organic matter and from the metabolic wastes of animals. Organic-based materials certainly improve soil structure, as well as providing nutrients to plants and microorganisms in the soil, and Table 3.3 briefly lists some of the different types.

You should note that the composition of compost varies, with some commercial types being of poor quality, but some are great.

Table 3.3 Organic-based soil amendments

Common name	Composition	Uses
Compost	Decayed plant and animal residues. Hot composted method is best as pathogens are killed.	General plant fertiliser. Increases organic matter and humus in soils. Add to all soils.
Wood ash	Ash from wood burning contains potassium, calcium, phosphorus and magnesium.	Wood ash is 8% potash – potassium compounds. Reduces acidity, repels pests. Alkaline.
Charcoal, biochar (Carbon)	Made from burning plant matter in little or no oxygen at a high temperature.	Source of carbon for micro-organisms in soil. Slow to decompose, long-lasting.
Mulch	Wood pieces, shredded plant material – leaves, stems, roots.	Surface cover to reduce water loss from soil, prevents overheating. Some nutrients available to plants when broken down.
Blood and bone (meat, blood, bone meal)	Waste products from abattoirs. High in nitrogen, calcium and phosphorus.	Slow release fertiliser, providing gentle long-term feeding of plants and soil.
Straw	Most often cereal plant stalks after seed-heads are harvested. Dry plant material.	High in carbon, food source for microbes. Improves water retention.
Compost tea	Aerated solution of microbes and nutrients made from plant material or manures, or both.	Spray over soil to add microbes and nutrients.
Manure	Aged or rotted animal manure from stock, poultry or wild animals.	High nitrogen and organic matter content. Do not use fresh as may burn plants.
Biofertiliser	Cultures of living organisms such as nitrogen-fixing and other bacteria, fungi. Often made as an anaerobic solution by fermentation.	Addition of soil biota to enhance processes.
Seaweed extracts, fish hydrolysates	By-product from fish processing or seaweed fermentation.	Source of wide range of plant nutrients to increase crop performance.
Vermicasts (earthworm extracts)	Worm castings used as produced or made into a solution.	Good source of nutrients and humified organic matter.
Humic and fulvic substances	Materials extracted from compost, peat or other organic matter.	Liquid form used as foliar sprays or solid form as granules into soil.

Biochar

A special mention should be made about biochar as this is gaining popularity in farming circles. In general, biochar is produced by burning wood or other organic matter (such as straw and manure) at a high temperature and low oxygen environment where volatile vapours are released leaving behind a solid mass.

When organic matter is burnt without oxygen or very limited oxygen, the result is a lump of black biochar rather than a piece of wood or other material being burnt so all that remains is ash. Charcoal is also made in a similar way. Biochar can be thought of as a form of charcoal but there are some small differences. Charcoal is burnt slowly and at low temperatures so it contains tars and other substances, and is used as a fuel. Biochar is made at much higher temperatures, resulting in carbon with lots of pores (holes) so it is light (less dense), and it is used as a soil amendment.

Biochar is essentially a block of carbon with small amounts of other elements, such as hydrogen, nitrogen and various trace elements scattered throughout its mass. Biochar might contain 80% carbon and 3% nitrogen by weight, but these amounts vary with temperature. For example, when making biochar at low temperatures (below 200°C) proportionally more nitrogen is found, but at higher temperatures (500°C and higher) the proportion of carbon is greater as nitrogen tends to be lost or converted to complex compounds which are held in the biochar mass.

The process of biochar production recycles plant waste which is made into a useful amendment and fertiliser. The porosity and large surface area of biochar is effective at holding nutrients including nitrogen compounds and inhibiting their loss through leaching. Ammonium and nitrate ions can be absorbed by the biochar and the pores are also sites for soil bacteria to act on these nitrogen compounds as part of the nitrogen cycle.

The physical and chemical characteristics of biochar is greatly influenced by the heating conditions used to make it and by the type of organic matter used in the process. Biochar might have different pH levels (be slightly acid to alkaline – pH 6 to 10), have varying carbon to nitrogen ratios (C:N could be 30:1 to 50:1 as an example) and different ash fractions. Biochar made from wood has higher C:N ratios than that made from grass, and biochar particles become smaller as the pyrolysis (heating) temperatures increase. The addition of alkaline biochar to acidic soils, common in farming country, can alter the pH levels yielding greater production in crops and a wider range of nutrients available to them.

In summary, biochar can be a useful amendment to add to soil. Biochar is very slow to decompose so it locks up carbon in soils. The organic matter in biochar is stable and while soil carbon can be decomposed and returned to the atmosphere, the process takes a very long time. Biochar has been known to increase soil water-holding capacity, increase soil microbe biomass, absorb a range of nutrients, and reduce carbon mineralisation (loss of carbon via carbon dioxide to the atmosphere) in the long-term.

Troubleshooting your Soil

Table 3.4 lists the types of amendments to use to change particular soils. In the tables listing amendments, peat isn't discussed. Peat is found in several forms but if it is peat from former swamps or peat bogs (partially decayed, high in carbon) then this is not a sustainable product. It is a limited non-renewable resource.

Table 3.4 What amendments to use for difficult soils

Soil problem	Amendment
Too much clay	Gypsum, coarse organic mulch
Too much sand	Bentonite, compost
Acidic	Limestone, lime or dolomite, wood ash
Alkaline	Sulphur
Poor water retention	Zeolite, compost or bentonite
Poor nutrition – few nutrients	Rock dust, zeolite, potash, rock phosphate, compost

Sphagnum moss, another form of peat, is usually harvested above ancient peat beds and in some cases harvested sustainably so more can grow. These are small moss plants which are harvested, dried and compressed. However, it may be better to source coir peat which is a peat-like by-product from coconut fibre.

Biostimulants

Biostimulants are human-made or human-assembled plant exudates. A range of substances are manufactured and applied to plant leaves, plant roots and to the soil to enhance plant growth, microbe activity and plant resistance to stress, disease and pest attack.

Common biostimulants include seaweed extracts, humic and fulvic acids, liquid manure or decomposed plant solutions and beneficial bacteria and fungi (e.g. rhizobium and mycorrhiza) granules. Milk is a biostimulant which contains calcium, sugars, and various vitamins, minerals and proteins to provide a quick feed boost to both plants and microorganisms. These substances increase microbial activity and then the microbes can make nutrients more readily available to plants. However, their effectiveness is variable and depends on the soil temperature and moisture content, organic matter content and weather conditions on the day of application. Some products may not have any real effect on the agricultural system, so it is best to trial the products and observe the outcomes before large scale adoption is undertaken.

When farmers are consciously decreasing herbicide use during their transition to regenerative agriculture some may add humic or fulvic acids to the herbicides to enable crops to recover and not be set back.

Covering the Soil

Covering the soil is an important regenerative agriculture principle. Soil cover, which may include cover crops, perennial grasses and crops, mulch and green manures, is discussed in a later chapter, p.98.

While various mulches do reduce water loss and the overheating of soils, the use of plastic sheet to cover soil is not advisable. The sun passes through the plastic and the heat build-up (solarisation), cooks the soil to kill disease and pests, but unfortunately also kills beneficials. Bare soil also experiences temperatures to 65°C, and at, and beyond, this temperature all life ceases to exist. Mulches should be ideally made from once-living things, such as plant prunings, chipped wood, sawdust or straw.

Soil Management

A farmer doesn't need to be a soil scientist to recognise the importance of soil biology and soil health as crucial to plant growth, plant nutrition and food nutrient density. Sound soil management is required if any changes to soil quality, soil properties or soil organism types or numbers occur, as these influences can hinder optimum soil functions and soil complexity. The soil food web interactions are crucial for high soil productivity and these need to be protected and maintained.

Farmers need to recognise the importance of taking regular soil tests to monitor either improvements or decline in soil fertility and structure. Soil tests help to ascertain what you have to start with or what it has degraded to. A comprehensive suite of mineral analysis is ideal, as well as the Haney test to determine the C:N ratio and the amount of organic matter available for microorganisms and macroorganisms to feed on.

Soils also have a number of ecosystem services which include carbon sequestration, water purification, nutrient recycling, habitat for organisms, water movement and flood regulation, provision of materials for housing and buildings and, of course, provision of food, fibre and fuel. The soil, in one way or another, enables 95% of our food to be made available to us.

You can see why soil is so important, why it needs protection and why it has to be regenerated. To best manage soil we need to concentrate on:

· Land preparation – which includes tillage or no tillage, contour ploughing and keyline planting;

· Water use – how much water is stored in the soil or accessed from water harvesting strategies;

· The type of crop or seed we select and use in terms of growth habits and tolerance to environmental change;

· Various agricultural inputs such as biofertilisers, manures and amendments; and

· All of the tools we can use to instigate good practice.

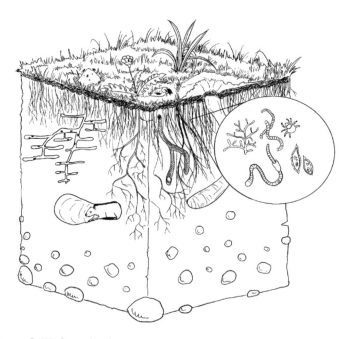

Figure 3.17 Sound soil management will ensure high productivity.

A healthy farm relies on soil health, environmental health and economic health. Healthy farm practices may include crop diversity, crop rotation, biodiversity, cover crops, permanent sod, integration of crops and livestock, and soil protection strategies. This is becoming increasingly important as farmers experience unreliable seasons, when crop production and yield shifts from extremely poor to a bumper crop.

We have to really get the soil right (with all of the important microorganisms present) if we want to wean ourselves off synthetic fertilisers. Anything we can do to reduce input costs will be beneficial to farmers. The transition to low chemical farming may take several years, but you could expect to see some gains after the first two years and significant changes within five years. After 10 years you could expect the production average and profit to be consistently higher before transition occurred.

We will discuss more about soil and its relevance to agriculture in the following chapters.

4

All about Plant Function

Before we delve into the different types of plants that can be grown in a regenerative agricultural system, we need to first discuss some of their features and functions, so you can appreciate how they relate to soil, other organisms and the environment. Useful plants in a regenerative agricultural system is discussed in the next chapter, p.85.

Processes in Plants

Plants are living organisms. Like all living organisms, plants respire, grow, excrete, move, respond to stimuli and reproduce. However, they are distinct from animals because they can photosynthesise. Photosynthesis is the key to life on Earth. All life directly or indirectly relies on photosynthesis.

Photosynthesis is the process that occurs in plant leaves where simple substances such as carbon dioxide and water are changed into sugars and oxygen. Plants are able to capture sunlight and store this as chemical energy in the myriad of organic substances they make. The initial sugars are then changed into many other substances which both plants and animals use. What is important though, is that plants produce about 10 times more oxygen than they consume in a day through respiration. This means that plants provide the oxygen and food that all other organisms need. You could think of a plant as a solar powered carbon farm.

Photosynthesis relies on the sun. It is the sun's light that provides the energy chloroplasts need to combine carbon dioxide and water. Seems simple enough, but scientists cannot make sugar and oxygen from any number of experiments they have tried over many decades. The process of photosynthesis has many steps and many other substances are involved but, even so, photosynthesis is one of the most remarkable processes that nature provides.

In essence, plants are energy fixers, using sunlight to make chemicals, and we want this to occur every single day of the year. However, in many places in the world there are dry seasons with no rain to maintain grasses and other plants. We will look at this later in the chapter (p.69) and when we discuss different types of plants, in the next chapter (p.96), which overcome some of these climate limitations.

Respiration, in terms of reactants and products, is the exact opposite of photosynthesis. In respiration, glucose or other sugars and substances react in the presence of oxygen to produce carbon dioxide and water. These latter two substances are essentially waste products and are released, emitted or passed out of organisms. Respiration that occurs in the presence of oxygen is called aerobic respiration. Like photosynthesis, respiration occurs in steps and needs other chemical substances to make the process work. However, there are variations to what happens under particular circumstances. For example, many bacteria live in low oxygen conditions, such as those found in soil or under water. They undertake anaerobic respiration, which is respiration without oxygen. This process is commonly known as fermentation. Here enzymes react with glucose and produce carbon dioxide and substances such as lactic acid or ethanol.

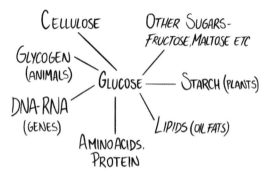

Figure 4.1 Products from photosynthesis.

Once the glucose is made in the leaves it is often converted into sucrose which is the main transported carbohydrate sent to other regions of the plant. Sucrose is table sugar and is found in high amounts in sugar cane and sugar beets. In cells, some glucose is used in cellular respiration and some is converted to many other substances and used in other processes. Besides sucrose, plants convert glucose into starch and this is stored in organs for later use. Potatoes, carrots and many other plant tissues store starch. The movement of glucose, sucrose and other organic substances throughout the plant is called translocation.

About 95% of a plant biomass is made up from substances originally made from oxygen and water then changed into a myriad of organic compounds: The various molecules of life. While glucose is the initial compound it is changed into many others, such as: Fructose (fruit sugar), sucrose (cane sugar), maltose (malt sugar), lactose (milk sugar), cellulose (the structural component of wood, stems and leaves), starch (storage compound in plants), glycogen (storage in animals), oils and fats (lipids), and combined with nitrogen (sometimes sulphur too) and made into amino acids, proteins and nucleic acids (DNA and RNA). The storage compounds, such as starch, are polysaccharides which means they are composed of many sugar molecules joined together in long chains.

About 25-30% of wood is composed of lignin. Lignin is a complex polymer made from chains of cyclic compounds. It is also found as structural components of plant vessels which transport water and sugars throughout the plant. Another 40-50% of wood is made from cellulose.

We can measure the amount of glucose, sucrose and minerals in plant tissue and this indicates the nutritional value of the food, the nutrient density and its general health. One simple apparatus to do this is the Brix refractometer. You use a garlic crusher to extract a few drops from leaves or soft stems so that measurements can be made, and it is no use measuring things if you don't record things. Drops of sap are placed on a surface and the device is pointed towards the sun (light). A direct reading on the visual scale indicates the sugar content. The measured result is called the degree Brix and this is converted to a percentage of sugar and displayed as % Brix.

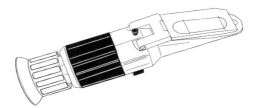

Figure 4.2 Refractometer.

One degree Brix is 1 gram of sucrose in 100 grams of solution and represents the strength of the solution as percentage by mass. If the solution contains dissolved solids other than pure sucrose, then the degree Brix (°Bx) only approximates the dissolved solid content.

Brix can be used to measure sucrose found within vegetables, juices, fruits, beer, and many more plant-based foods. We use these readings to examine our food crops as there is a direct correlation between a plant's Brix value and its taste, quality, potential alcohol content, and nutritional density.

Generally, crops with a higher refractive index (Brix value) will have a higher sugar content, higher protein content, higher mineral content and a greater specific gravity or density. This might add up to a sweeter tasting, more nutritious mineral feed and better storage attributes. We also associate a high reading with high soil fertility, pest resistance and resilience, so using these simple tools in the field helps the farmer to determine how well the crop is going. Healthy plants with high Brix are more effective at resisting pests and plants with Brix greater than 12 have been shown to have better frost tolerance as the sugary sap succumbs to less frost damage.

Examples of the range of Brix percentage readings are shown in the table that follows. Any readings below the 'average' is deemed to be poor.

The other major plant process relates to water movement. This process is called transpiration and it is the movement of water from the soil through roots into the stems and leaves and finally out through the stomata pores where water vapour is released to the atmosphere.

The continual stream of water upwards carries mineral ions to the leaves. Various minerals are needed for the many cell reactions and the various compounds that are made in plant leaves. When not enough water is in the soil, plant leaves wilt and sometimes the plant is beyond saving and will die.

Plants also transport sugars and other organic substances downwards, in special vessels, to other plant parts such as stems, tubers and roots. This fluid is the sap that we measure with our refractometer.

More trees means more transpiration, more clouds and ultimately more rainfall. As the amount of rainfall drops the run-off reduces at a greater rate – not exponentially but a greater consequence.

Table 4.1 Brix percentage readings in some plants

	Average	Good	Excellent
FRUITS			
Apples	10	14	18
Cantaloupe/rockmelon	12	14	16
Grapes	12	16	20
Mangos	6	10	14
Oranges	10	16	20
Strawberries	10	14	16
Tomatoes	6	8	12
GRASSES			
Alfalfa	8	16	22
Grains	10	14	18
Sorghum	10	22	30
VEGETABLES			
Bell peppers/capsicum	6	8	12
Broccoli	8	10	12
Carrots	6	12	18
Peas	10	12	14
Potatoes	5	7	8
Sweet potatoes	8	10	14
Sweet corn	10	18	24
Turnips	6	8	10

Finally, plants function to produce seeds which make more plants. While not examining how all of this occurs, the process is again reliant on the substances that plants make. In this case, various hormones are involved.

Two hormones found in seeds determine germination: Abscisic acid and gibberellic acid. Abscisic acid makes seeds dormant while gibberellic acid make seeds germinate. Various triggers affect the levels of these hormones. These might include sugars (glucose, sucrose), nutrients (nitrate, nitric oxide) and the carbon:nitrogen ratio critical for the formation of nitrates. Furthermore, when seeds are coated with bacteria and fungi it often ensures a high germination rate and successful establishment. Then, when plants flower (and seed) you get a lot more pollinators such as insects and birds in the system. You can see how intricate all of the processes in plants are, and how complex it sometimes is.

Cool and Warm Season Plants

While animals eat all year round, there are no plants that can be used as forage all year round. Forage are those plants that are grazed. Grazed plants respond by photosynthesising more and producing more carbon compounds.

Some plants die back in cooler months, others suffer in high-temperature environments. Knowing that some plants are called C3 (cool season, temperate) and some plants are referred to as C4 (warm season, tropical) is the key to having quality forage all year long. C3 and C4 plants are called that because of their different metabolic pathways when they photosynthesise and make sugars.

C3 and C4 plants differ in their leaf structure and the enzymes they have that carry out the photosynthesis process. In photosynthesis, carbon dioxide is changed, in a series of steps, into a long chain of six carbon atoms to make a simple sugar. C3 plants have a particular enzyme that creates a three-carbon acid (phosphoglyceric acid) which eventually forms the six-carbon chain of glucose (a hexose sugar). C4 plants, on the other hand, as you might expect, change carbon dioxide into a four-carbon acid, oxaloacetate.

There are lots of other differences between these two groups of plants. C3 plants fix carbon dioxide more efficiently in cooler months, produce a higher percentage of protein and usually have higher forage quality and nutrition as these plants can be digested faster, but their growth is reduced in summer. Examples of C3 plants include wheat, rye, tall fescue and oats.

C4 plants tend to have higher efficiency in carbon dioxide uptake and nitrogen use, use less water to make dry matter, have more fibre, and grow and are more active in warmer months. Examples of C4 plants include corn, pearl millet, Rhodes grass, green panic, paspalum, kangaroo grass, Bermuda grass, kikuyu and sorghum. Seasonal growth patterns of both C3 and C4 plants are shown in Figure 4.3, while Figure 4.4 shows how a variety of different plants can provide forage all year round.

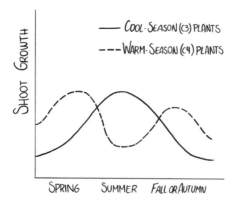

Figure 4.3 Seasonal growth patterns for C3 and C4 plants.

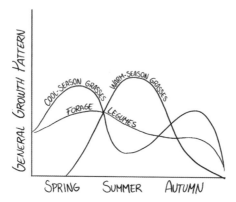

Figure 4.4 An integrated forage system.

Some 93% of plants are C3, only 1% are C4, and the balance (6-7%) are CAM – Crassulacean Acid Metabolism plants – which are those plants that thrive in deserts and dry environments. Cacti and pineapples are examples of CAM plants.

CAM plants allow carbon dioxide gas to enter the plant at night when it is cooler. It is held there and released during the day inside the leaf so photosynthesis can occur. The stomata are closed at night to prevent or minimise water loss. Plants that can survive in regions where there is little liquid water, such as deserts or in ice or snow environments, are called xerophytes.

Table 4.2 compares C3 and C4 plants, but this is a general comparison as there is a wide variation in plant species within each group. However, it is a useful guide to help farmers make decisions about what kinds of plants they should be growing. In most scenarios, some combination of C3 and C4 plants should enable properties to provide forage and fodder, as well as ground cover, all year round. You might use C4 grasses, which slow down and are dormant in winter, and then C3 cash crop cereals can be sown so they get up above the perennials and vice versa. However, don't be disappointed when some species die out. Much of your journey will be trial and error, so if something doesn't survive or work, try different plant combinations or different approaches.

They say there is a difference between edible and palatable. Some wild foods we come across may be edible (we won't die) but they may not be palatable (we won't enjoy it). This means that if we have a choice or know what particular foods taste like, we may decide not to consume them. Animals are the same: They prefer some feed, grasses and shrubs over others. Palatability is about having a preference for a food if we have a choice.

Table 4.2 A comparison between C3 and C4 plants

Characteristic	Cool season plant Tropical	Temperate Other examples
Warm season plant	Soybean, barley, chard, broccoli, beetroot, clover	Switchgrass, sugarcane, lablab, pigeon pea, amaranth
Metabolic pathway	C3 (3-carbon acid)	C4 (4-carbon acid)
Above-ground biomass/yield kg/ha	Lower	Higher
Lifecycle	Annual or perennial	Annual or perennial
Seasonal growth	High in winter to spring	High in summer
Nutritional value	Higher	Lower
Palatability	High in cooler months, declines in summer	High in summer, declines with age
Frost tolerance	Good	Poor
Digestibility	Higher	Lower
Crude protein	Higher	Lower
Fibre	Lower	Higher
Light requirements	Lower	Higher
Temperature requirements	Lower	Higher
Nitrogen requirement	Higher	Lower
Salinity tolerance	Lower, poor	Higher, good
Shade tolerance	Good	Poor
Photosynthesis efficiency	Lower	Higher
Water requirements and content	Higher	Lower
(General) overall size	Shorter	Taller
Fructan (storage carbohydrate)	Present	Absent
Structural carbohydrates	Higher	Lower
Starch (as storage carbohydrate)	Lower	Higher

Plants are more palatable to animals when young, tender and leafy. As the plant stems thicken and mature, the roughage and woody nature becomes less palatable. As animals prefer young plants, as both palatability and digestibility are better, it is important that plants have enough time to recover after being eaten and grow new leaves, which are the sites of photosynthesis and where products are made to promote growth.

There is always a balance to when stock are allowed to graze. If the grass and shrubs are grazed early, yield can be sacrificed and the plant community weakened. If the plants are allowed to mature then the feed quality may be poor and weight gain in stock may fall. The most important thing is to endeavour to have green plants all year round, so again, some combination of cool season and warm season plants, both annuals and perennials, is ideal in most soils and climates.

Besides C3, C4 and CAM plants all having different metabolic pathways, they respond differently to environmental conditions. Some plants we call pioneers, others are climax, and the interplay between plants and their environment is succession, and this topic is discussed in the next chapter, p.85.

Nutrients for Plants

Nutrients are those substances that every living organism needs to survive, grow and develop. There are about 118 known natural and human-made elements on earth, but only about two dozen of these are required by all life. Nutrients must be available not only in sufficient amounts but also in appropriate ratios. We often don't really know what nutrients a plant needs and we have to rely on the microbes that supply the plants.

Plant nutrition is a difficult topic to generalise, simply because different plants require different amounts of nutrients. There are about 16 or 17 elements that are essential nutrients for plants (although about 20 elements are required by plants in one way or another), with those required in high amounts called macronutrients and those only needed in small amounts as micronutrients or trace elements.

Compounding all of the issues about nutrient availability is that if the balance of nutrients isn't right, then toxicity and deficiency looms, and sometimes too much of one element causes the reduced uptake of another element. We don't normally talk about carbon, hydrogen and oxygen as macronutrients because these are the backbone of organic substances and most compounds in plants contain all three. We simply say these are the basic elements as they are supplied by carbon dioxide and water.

While some macronutrients are required in larger amounts than others (the primary macronutrients are nitrogen, phosphorus and potassium) no distinction is made here. All elements have a role to play in plants and Table 4.3 is a brief look at some of them.

Table 4.3 Common nutrients required by plants

Element	Functions
Nitrogen	Found in every cell and many organic compounds such as chlorophyll, amino acids and peptides. Nitrogen is typically sourced from the soil or through nitrogen-fixing bacteria.
Phosphorus	Phosphorus is also found in many compounds in plants and these include nucleic acids (DNA and RNA), phospholipids that make cell membranes and adenosine triphosphate (ATP) which is the immediate source of energy for cells. The process of photosynthesis and respiration both rely on ATP to provide the energy for the various reaction steps. Phosphorus is decreasing worldwide, so we need to make better use of what is available in the soil already.
Potassium	Potassium is unique amongst the macronutrients. It is not used as a structural component of any organic compounds, but is essential for enzyme activity and to regulate the opening and closing of the stomata in leaves. The stomata are tiny holes in the leaf surface where gases (oxygen and carbon dioxide) can enter and leave the plant.
Sulphur	Sulphur is found in many amino acids and vitamins and is needed for part of the nitrogen-fixation process. Sulphur can be used as an amendment to make alkaline soil more acidic.
Calcium	Calcium is chiefly used to build cell walls, giving the plant rigidity and stability. It has roles in root development, preventing disease, and cell division. Calcium compounds such as calcium carbonate (limestone), dolomite (a mixture of calcium and magnesium carbonates) and calcium oxide (lime) are all used as soil amendments to make acidic soil more alkaline. Of all the minerals in the soil, calcium is the key to soil recovery. Besides animals and humans requiring calcium for bones and growth, plants use calcium in many ways.
Magnesium	Magnesium is the mineral component of chlorophyll and supports enzyme reactions. Many soils are deficient in magnesium, so dolomite is a useful amendment to add to depleted soils.
Iron	Iron is used as an enzyme cofactor to make chlorophyll so iron deficiency often results in yellow leaves. Iron is not usually low in most soils, and soils that are yellow, orange or brown all contain iron – that's why they have those colours.
Boron	Boron is an element that is unfamiliar to most people. However, it has many functions in a plant: it affects flowering and fruiting, pollen germination and the metabolism of many compounds.
Copper	Copper is important for photosynthesis and to produce lignin, a structural component of plant cell walls.
Manganese	Manganese is involved in making chloroplasts which is where photosynthesis occurs, so any deficiency often results in light spots in a leaf.
Sodium	Sodium is mainly required in C4 rather than C3 plants and is important for water balance. Too much, however, causes plant stress and in clayey soils may cause dispersion.
Zinc	Zinc is another enzyme co-factor and pays an important role in many cell reactions. Like all of the last few elements listed above, zinc is a trace element and is only required in small amounts.

Other elements that plants require (still classed as trace elements or micronutrients) include nickel, chlorine, cobalt, silicon, molybdenum, selenium and iodine, but these are not discussed here. While all of these are not essential for survival they do contribute to plant growth and development, as well as being important in our diets too.

The other thing to consider is that many herbicides are chelators, binding with metals and locking them in complexes so they are not available to plants, and therefore cannot help plants with metabolism, fighting disease and other plant functional processes.

Nutrient (Element) Interaction

Many elements interact and influence their uptake, or not, by plants. Dutch scientist Derek Mulder published a paper in 1953 ('Minor Elements in Fruit Growing' – English translation) that showed how different elements interacted. Some hindered nutrient uptake, and some enhanced uptake. Figure 4.5 shows one variation of his chart. Look at the direction the way the arrows are pointing and in what direction. Those elements that interfere with one another are said to be antagonistic, while those elements that have a positive influence are stimulants. The processes involved are thus antagonism and stimulation.

As examples, high levels of calcium can reduce the availability of manganese, boron, potassium, iron and zinc, while high levels of nitrogen can create a demand for magnesium (the result could be a magnesium deficiency) and high levels of molybdenum cause an increase in the need for copper. As copper is often deficient in soils or forage, it causes deficiencies in animals and they may need supplements added to their diet. Calcium and phosphorus are synergistic as they have a mutual effect on each other: Increasing one increases the absorption of the other, and vice versa.

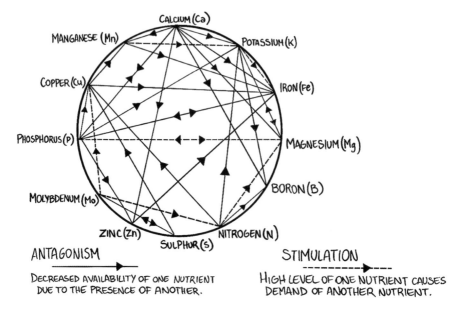

Figure 4.5 One variation of Mulder's Chart.

It doesn't matter if the process is antagonism or stimulation, the result is much the same: Induced deficiencies if the crop is not given a balanced supply of nutrients. And that is the key, make sure that all plants are given a broad suite of nutrients so that all of their needs can be met. Don't just focus on NPK and macronutrients, but think about supplying a good range of trace elements too. Maybe we should replace the old NPK with the new NPK – New skills, Practice and Knowledge.

Mulder's Chart can be a useful guide to enable farmers to understand why their plants look sickly or yellow or withered. However, there are also other considerations. Many nutrients are supplied to plants by mycorrhiza fungi and through the actions of bacteria, so element interactions can be complex and not as straightforward as we often hope.

Nutrient Deficiency in Plants

It is important to get the nutrition right in both plants and animals so they both can have optimum growth, health and wellbeing. We can usually recognise nutrient deficiencies in plants by examining their leaves. While Figure 4.6 highlights typical symptoms, sometimes these change a little from plant to plant. Furthermore, symptoms due to disease (bacterial, fungal or viral infections) aren't shown here and some of these can have similar symptoms to nutrient deficiencies. The main issue to be aware of is that both poor and over nutrition (fertiliser) on plants reduces a wide range of important substances in plants, and reduces their production capacity.

Furthermore, you don't normally see nutrient deficiency in natural forested areas, so it appears that our agricultural practices are causing plant and soil problems. To help us determine possible deficiencies (or excesses) then leaf tissue tests, soil tests and plain observations are required. It is important to get a suite of results for many minerals so you know what deficiencies there may be, what to add and how to correct and ameliorate problems.

Once we have the information we are able to apply the proper treatment. If anything, foliar sprays onto leaves are much more effective in getting the nutrients into a plant. Solid fertiliser pellets could remain as pellets for some time and there is no guarantee any of the dissolved fertiliser will come into contact with roots and be absorbed. A foliar spray could also drip onto the soil surface and as it moves into the soil it is more likely to become assimilated.

The application of a foliar spray can be made as a transitional phase because it does provide a kick to growing plants. The long-term aim, however, is the better use of cover crops, soil management and the occasional use of non-toxic soil amendments.

As artificial fertilisers tend to be soluble, much is lost through leaching and pollutes the groundwater and other waterways. Some farmers also apply more than what is required simply to account for what is lost. This just adds extra

CALCIUM:

New leaves misshapen or stunted.
Existing leaves remain green.

IRON: Young leaves are
yellow/white, whith green veins.
Mature leaves are normal.

POTASSIUM:
Yellowing at tips and edges,
especially in youngs leaves.
Dead or yellow patches or spots
develop on leaves.

NITROGEN:
Upper leaves
Light green. Lower leaves
yellow. Bottom (older leaves)
yellow and shrivelled.

SULPHUR:
Leaves light green.
Veins pale green. No spots.

MANGANESE:
Yellow spots and/or
elongated holes
between veins.

PHOSPHATE:
Leaves darker than normal.
loss of leaves.

MAGNESIUM:
Lower leaves turn yellow
from inwards.
Veins remain green.

Figure 4.6 Symptoms of nutrient deficiencies.

input costs. Applying any additional fertiliser has to be seen as a short-term fix
and not the solution.

Bear in mind that the problems with applying superphosphate and other
similar phosphorus fertilisers is that phosphorus is easily absorbed by clay
particles as well as binding with iron, aluminium and calcium, so it is locked
up and not available to plants. Besides this, superphosphate is also acidic and
often contains heavy metals. By letting soil organisms do their magic, bound
phosphorus can be released. Many types of bacteria and fungi can mobilise
phosphorus, and we need to let nature do the work for us.

Effect of Acidity on Nutrient Availability

Nutrients are also influenced by acidic or alkaline conditions. Figure 4.7 shows
how the pH affects nutrient availability. The thick parts of each element line
suggest good availability while tapering lines suggest inhibition, which leads to
deficiencies. For example, in acidic conditions potassium and phosphorus are
locked up and not available while in alkaline conditions iron and manganese
may become deficient. Most elements are typically available at pH of about 6.5
to 7 (the soil chapter (p.27) provides more elaboration of acidity and pH). In the

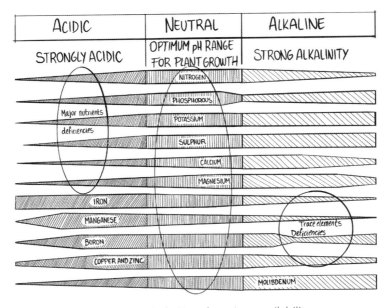

Figure 4.7 Soil pH and nutrient availability.

middle of the graph, most elements have thick lines around the neutral range (pH = 7), so we should aim to change our soils to be slightly acidic to neutral.

There is also another side to the effect of acidity on nutrient availability. For example, in acidic conditions iron is readily available but this can be toxic to plants, while in alkaline (basic) conditions too much molybdenum may cause stunted seedling development. Once again, it is always about balance – an appropriate acidity level in our soils to ensure all elements are available to plants.

Redox

There is a lot more to plants and soil than just fertilisers and nutrients and the chemistry that is involved in all of this. To be successful regenerative agricultural farmers and producers we will need to rely a lot more on our understanding of biology and biophysics. Biophysics is about understanding the mechanisms of how biological systems work. This branch of science looks at how plant cells capture light and transform it into energy, how stress affects plant function and how the molecules of life are created.

Besides acidity and element interactions affecting nutrient availability, other processes are in play. While we have briefly discussed respiration and photosynthesis, these reactions only take place when electrons are transferred as part of oxidation and reduction, or redox, reactions.

Oxidation is a process where electrons are lost from an atom, and in our early understanding of redox chemical reactions scientists described oxidation as adding oxygen to another element. For example, when carbon is burnt in air it changes into carbon dioxide: The carbon has been oxidised. In many early studies oxidation and reduction involved the gain of oxygen and the loss of hydrogen.

Oxidation never occurs without its opposite partner, reduction. Reduction is the process of gaining electrons and in our previous example when carbon dioxide is made, the oxygen gas is reduced because it is now part of another molecule. We now know that redox reactions don't always involve oxygen, or hydrogen, so we just talk about the loss and gain of electrons. An oxidation reaction strips an electron from an atom and the addition of that electron to another atom is a reduction reaction. You can remember what oxidation and reduction mean by the mnemonic OIL RIG: Oxidation is loss, Reduction is gain.

How important or relevant is redox reactions in plants? Extremely important, because redox reactions are how respiration and photosynthesis occur, as well as a vast number of other reactions in plant cells for protein function, gene expression, signalling and communication in plants, plant development and defence, enzyme reactions and the action of hormones and antioxidants.

Revisiting respiration and photosynthesis again, respiration occurs when glucose is oxidised to carbon dioxide and oxygen is reduced to water. During photosynthesis, plants take in carbon dioxide and water to make glucose and oxygen. The water is oxidised, meaning it loses electrons, and the carbon dioxide is reduced, meaning it gains electrons.

The redox potential (symbol Eh) is a measurement of the availability of electrons. It is measured in millivolts (mV), as electricity is flowing electrons, and is normally found to be less than 1000 mV or 1 volt in soils and cells. Redox is often associated and intertwined with acidity and pH. They are different and independent, but both are drivers of soil, plant and microbe systems.

The redox potential affects the availability of many nutrients. This is because the soil conditions alter the state of element ions. For example, the form of nitrogen can change when the redox potential changes from low to high, with ammonium favoured at low Eh and nitrate if the Eh is higher. Manganese is more bioavailable when the redox potential is lowered but it can be toxic in low acidity. High redox potential causes iron deficiency and the redox potential decreases when soils are waterlogged because the amount of oxygen decreases.

Figure 4.8 shows a Pourbaix diagram which illustrates how redox potential and pH affect various minerals. As an example of how to read and understand this diagram, under acidic conditions and an Eh of about 400 mV nitrogen tends to exist as ammonium, phosphorus and molybdenum may become deficient but aluminium may develop toxic levels. In an alkaline environment (pH > 9) and an oxidised state (Eh = 600 mV) then nitrogen tends to exist as nitrate and some trace elements, such as manganese and zinc, may become deficient. The most favourable conditions for plant growth and mineral availability is at pH about 6.5 to 7 and redox potential about 400 mV (0.4V).

As you can see, element availability, chemical reactions and plant development is complex, and we all need to find out more about redox potential as a way to correct poor soils, improve plant health and overcome mineral toxicity and deficiency.

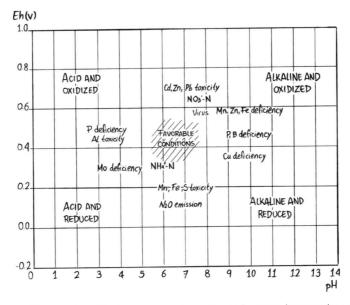

Figure 4.8 The relationship between redox potential, pH and mineral availability.

Nitrogen Fixation

A brief mention about the nitrogen cycle and nitrogen fixation as part of that cycle, was discussed in the soil chapter (p.43). Here, we focus a little more about this process and the types of plants that are involved.

Plants need nitrogen. As mentioned in the soils chapter, while nitrogen is very common in air, it is not available in this gaseous form to plants. It needs to be changed. Nitrogen is taken up by plants as nitrate and ammonium, and no plants can absorb and use nitrogen gas directly. Having oxygen in the soil drives the nitrogen cycle in ways that result in less nitrous oxide NO_2 being released to the atmosphere.

When nitrogen is low in soils it becomes a limiting factor to plant growth and productivity. One plant group, the legumes or pea family, overcomes this problem by developing a symbiotic relationship with particular bacteria. These bacteria live in swellings in the plant roots (nodules) and they have the ability to take nitrogen from the air and change it into ammonium and then into nitrate, both of which are soluble forms of nitrogen. Ammonium is a simple mixture of nitrogen and hydrogen (NH_{4+}) while nitrate is nitrogen combined with oxygen (NO_{3-}).

Other bacteria also act on these nitrogen compounds to make nitrites (NO_{2-}) and nitrogen gas (N_2), and all of these substances are part of the nitrogen cycle. However, nitrogen is easily lost from any system. About one-third of applied soluble nitrogen (nitrates) is not available to plants but is lost from the system, often by leaching (moving downwards bypassing roots) or bacterial action changing nitrates into nitrogen gas.

Ultimately, some forms of nitrogen are absorbed by plants and then passed on to animals. Nitrogen is the essential element found in all amino acids, proteins, deoxyribose and ribose nucleic acids (DNA and RNA), hormones (auxins) and many others including the toxic alkaloids. In plants, about 2% of the dry weight is nitrogen and about 40% is carbon.

Nitrogen fixation takes place in many different types of plants and many different types of bacteria are involved. For example, rhizobium bacteria are found in a symbiotic relationship inside the roots of peas, beans, clovers and alfalfa, azospirillum bacteria feed on plant exudates outside of the plant roots and rhodospirillum bacteria are free-living in the soil. Besides being nitrogen-fixing, rhodospirillum are phototrophic and when needed they can make their own food. The aquatic fern *Azolla pinnata* is the only fern that can fix nitrogen. It does so by virtue of a symbiotic association with a cyanobacterium (*Anabaena azollae*) living in the leaves.

In regenerative agricultural systems, planting legumes and other nitrogen-fixing plants as cover crops, companion plants, groundcovers, cash crops and for forage and fodder is a strategy to improve soils, improve productivity in food crops and keep the ground covered. More about these groups and types of plants is discussed in the next chapter, p.98.

However, it should be understood that legumes don't normally provide any nitrogen to other plants nearby in the field. The legumes themselves benefit from nitrogen-fixation but they don't share this with other plants. If the legumes were cut down or died then nitrogen could be released into the soil and bacteria there may capture and utilise this, with the possibility that some nitrogen is accessed by grasses and plants. It is a mistaken belief that legumes fertilise the soil with nitrogen or that nitrogen escapes from the legume roots into the soil so other plants will benefit.

Just because you plant legumes doesn't mean that they are even producing nitrogen through association with various nitrogen-fixing bacteria. Sometimes these bacteria are not present or in low numbers because the soil is just not right. Furthermore, particular bacteria only associate with particular plants and these need to be inoculated. On top of this is the fact that the nitrogen-fixing process is pH dependent – working much less efficiently in low (acidic) levels. Nitrogen-fixation also depends on the presence of metals such as iron and molybdenum which are part of special enzymes that bacteria require to help them in the process.

But, there is always a but. Some nitrogen can transfer to other plants by several indirect mechanisms. Mycorrhiza fungi can transfer nitrogen compounds through their network and from one plant to another, especially if different plants have roots close to each other and the fungi are attached to both plants. Animals, including stock, eat the plants and excrete manure and urine, both of which contain nitrogen compounds. About 80% of the ingested nitrogen is excreted in urine and faeces. Urea, for example, as the main excretory product in urine is 46% nitrogen. These nitrogenous wastes enter the soil and can be made available to plants. Finally, plant decay is another mechanism. Plants die and are decomposed, releasing nutrients back into the soil.

Common examples of nitrogen-fixing plants can be seen in Table 4.4. There is a mixture of cool season and warm season plants, some of which are used as cash or cover crops and others as forage or fodder.

Table 4.4 Nitrogen-fixing plants

Plant function	Cool season plants	Warm season plants
Cover or cash crops (or as green manure)	Field peas	Lentil
	Broad beans	Mung bean
	Alfalfa	Lablab bean
	Soybean	Sunn hemp
Fodder or forage	Medic (*Medicago* spp.)	Senna
	Clovers (red, white and others)	Trefoil
	Vetch	Cowpea
	Partridge pea	Blue lupins

As we discussed in Chapter three on soils (p.44), organisms need particular carbon to nitrogen (C:N) ratios to function and C:N also affects what plants grow. For example, if you just planted a legume, you may find in the next season grass growing and dominating in the paddock. To prevent these wild swings of carbon to nitrogen it is better to plant a legume with a grass in the mix. For example, field peas or clovers with oats or barley.

The C:N ratio in good soils is about 10:1. Usually carbon is in plentiful supply from decaying plant residue, and nitrogen can be limiting. If there are no animals on a farm we somehow need manure or other sources of nitrogen to supply the microbes to break down organic matter.

Finally, a brief word about the sulphur cycle because of its close association with nitrogen in many plant compounds. In the sulphur cycle, sulphur is reduced by bacteria to hydrogen sulphide or sulphide ions in waterlogged and compacted soils. So, in farms where these problems are common, sulphur is lost from soils. As sulphur is an important element in many proteins and amino acids, and considered an essential nutrient for plant growth (and all life as well), then having poor structure in farm soils exacerbates the problems farmers face when trying to build soil fertility. We need more of the other types of bacteria that change sulphur into the soluble sulfate which plants take up and utilise.

Rhizophagy Cycle

In the soil chapter (p.48) we examined the roles of bacteria and fungi in the soil and the interaction with plants. Here we take a closer look at the intricacies of microbe movement into and out of plant cells. This phenomenon is called the rhizophagy cycle – 'rhizo' means roots and 'phagy' is about eating, so literally it is about 'roots eating microbes' although this is not the actual case. Organisms that eat living microbes, such as bacteria, are known as microbivores, and microbivory is the process, which is common in animals but less so in plants.

In the case of plants, they do not 'eat' the microbes, but the microbes can be found inside living plant tissue. When organisms live inside other (host) organisms we call them endophytes. Endophytes include bacteria and fungi, such as mycorrhizal fungi. Endophytes have been shown to provide benefits to plants such as improved tolerance to environmental stress, increased disease protection and improved nutrient acquisition and growth promotion.

The association and relationships between plants and microbes, where both species benefit, is called symbiosis. Not all bacteria that have symbiosis with plants are nitrogen-fixing, but it is known that many types of bacteria contribute to increased plant growth, and more work needs to be done to fully understand these complex associations.

In the rhizophagy cycle, bacteria deliberately enter the root tips bringing nutrients from the soil with them. The plant cells provide oxygen-rich conditions which affect the microbes in different ways. Many are killed and the nutrients they harbour are absorbed into plant tissue. Bacteria contain the highest nutrient density in their bodies and these microbes are your plants' nutrient delivery guys. In effect, the plants are harvesting microbes and their nutrients.

The oxygen-rich condition initially comes from the oxygen which is continually produced in daylight hours by photosynthesis. Oxygen can be changed into a number of substances we call reactive oxygen species (ROS). 'Species' is probably not the best term as this is most often associated with names of organisms. It may be better to think about 'substances' so ROS could mean or imply reactive oxygen substances.

Reactive oxygen is most often oxygen ions (O_2-) and radicals (hydroxyl, OH-) which are both associated with hydrogen peroxide (H_2O_2) and these substances act on lipids (fats and oils), DNA and proteins. Reactive oxygen substances can also adversely affect the plants themselves, so plants produce a range of antioxidant compounds to keep the reactive oxygen levels in balance.

Some bacteria, however, seem to tolerate reactive oxygen substances and are not degraded by them, so nutrients are not released into the plant. Bacteria may be resistant to reactive oxygen substances due to their own production of antioxidants.

Although research is in its early days, the rhizophagy cycle appears to occur in all plants and may be an important way that plants acquire some nutrients.

However, the effect of bacteria on plants is two-fold: Some seem to enhance plant growth while some inhibit plant growth.

As we have discussed in the soil chapter, plants exude substances, such as sugars, organic acids and proteins, and these substances attract (and feed) bacteria in the soil. Plants are known to secrete exudates in nutrient-limiting soils, as the exudates act as signal molecules that attract a diverse community of microbes. When nutrients are scarce, plants increase the cultivation of microbes by increasing exudate production. The bacteria also absorb various nutrients from the soil themselves so they are, in effect, nutrient storehouses.

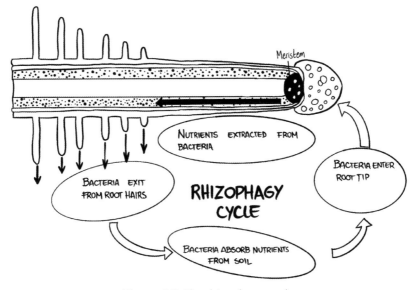

Figure 4.9 The rhizophagy cycle.

So plants cycle nutrients through exudated substances and decay. Research has shown that plants literally change the levels of nutrients in the soil around them. There are vast increases in organic matter, phosphorus, potassium and sulphur among other nutrients and pH may be ameliorated.

A large number of different groups of bacteria, that typically are found near or on plant roots, can be induced to enter through the root tips. This is because the root tips are actively growing and dividing, so they do not possess thick cell walls which are characteristic of all plant cells. Once inside the plant cells, the bacteria lose their own cell walls and are trapped there. The plant cells provide an oxygen-rich environment which affects the bacteria. The oxygen kills and breaks down some bacteria, so the nutrient stores they contain spill out and plants can use these. Those bacteria that survive make their way towards root hairs which are also thin-walled (as their cells elongate and grow), and they can exit back into the soil. Once outside the plant the bacteria reform their thick cell wells, absorb nutrients from the soil and the cycle continues.

Fungi also appear to have a role in the rhizophagy cycle. Several species of yeasts have been observed entering and exiting plant roots, so it is likely they would be exposed to reactive oxygen substances and degrade, releasing what nutrients they contain into the plant cell cytoplasm.

Nutrient exchange between bacteria and plants is well-known with the most common example being the variety of nitrogen-fixing bacteria living in plant root nodules and within other plant cells, but only recently have we found out that other soil nutrients are absorbed by plants through indirect pathways involving bacteria. Nitrogen-fixing bacteria don't perform that process inside plant root cells as the oxygen levels are too high and these types of bacteria thrive in low oxygen conditions. Key enzymes in the nitrogen-fixing process are inhibited by oxygen.

Nitrogen is not necessarily provided from fixation but from the digestion and assimilation of nitrogen compounds, which include amino acids and proteins, from other organisms. So bacteria can live inside plants and fix nitrogen or they can sequester nitrogen in the soil that surrounds plant roots. Bacteria (and fungi) have the ability to produce soluble forms of nutrients, such as nitrogen and phosphorus and these may be absorbed directly by root hairs.

However, plants need a whole host of nutrients and we are just starting to find out that other macro and micronutrients are also made available by the actions of bacteria and fungi. Rhizophagy might explain how boron, cobalt, iron, magnesium and zinc, for example, can be made available to plants when many of these nutrients are traditionally found in low amounts in the soil. More research needs to be undertaken to examine which nutrients are taken into plants by bacteria and how the plants actually get those nutrients from the bacteria.

By examining plant function and physiology a better grasp of the importance of their interaction with soil organisms and with nutrients is possible and an appreciation of our role in providing the optimum conditions for all of this to occur.

Final Remarks

Isn't nature amazing? We probably under-appreciate how complex nature is and sometimes the solution to a problem is not a simple fix. If plants aren't growing well this could be due to a range of problems such as erosion, compaction, nutrient imbalance, organic matter and carbon loss, pollution from chemicals and heavy metals, disease and disorders, acidification, acid sulphate and basic (alkaline) soils, low levels of beneficial fungi and bacteria, including mycorrhizal fungi and nitrogen-fixing bacteria, low or high redox potential, and salinisation and sodicity which relate to high levels of salt and sodium in soils.

Sure, we will make mistakes as we endeavour to learn more about these things and try different techniques, but nothing is gained unless we give it a go and work towards building and growing healthier soil and plants on our home, Planet Earth.

5

Using Plants in
Regenerative Systems

Every living organism relies on plants in some way, whether it is as simple as a food source, or just the oxygen we breathe. We need plants for medicines, fuel, windbreak, timber, honey, fodder, shade, mulch, fertiliser, cosmetics, cork, pest repellents, erosion control, oil, water purification, paper, shelter, chocolate, dyes, perfume, weaving materials, wax, rubber, rope, spices and clothing, and this list continues.

Before we examine how remarkable plants are, we need to understand a little more about the types of plants that exist, how plant communities change over time (succession) and the roles that plants play in the environment. At the end of the day, plants will create profit for the farmer provided they are managed.

Ecological Succession

Ecosystems are dynamic. They change as species move in and out, the types of plants change over time, and the soil also changes. All of this results in continual disturbance, which drives the nutrient cycles and interactions within the ecosystem. This continual change is called succession.

Ecological succession is a series of progressive changes in the species that make up a community over time. Succession often involves a progression from communities with lower species diversity, which may be less stable, to communities with higher species diversity, which may be more stable, though this is not a universal rule. Essentially, one community of organisms replaces another until a stable system is formed.

There are two types of ecological succession:

1. Primary: When organisms colonise new bare ground, such as after a volcanic eruption. The transition of bare rock to soil with early mosses and lichen giving way to small plants and then onto larger plants and then climax trees may take a thousand years. Certainly, when nature takes its course, it might take decades before the organic matter is high enough to allow those higher-level plants to colonise.

2. Secondary: This succession type occurs when organisms re-colonise a
   disturbed site. This may be after a severe fire or agricultural cultivation.
   Since there could be some organisms still present in the soil and landscape,
   a more rapid colonisation process takes place. Cultivating soil will have
   weeds at first rain, but the process of changing agricultural land back
   into 'nature or forest' may take 50 to 100 years. It might take 150 years to
   re-establish a forest after a severe fire. The damage we sometimes do to
   landscapes is not easily repaired by nature, certainly nature doesn't work
   fast and ecological repair and succession takes decades to generations.

The process of succession is relatively straightforward. Plant communities generally progress through predictable successional sequences. Particular plants are the first colonisers (pioneers) and then the system progressively becomes more complex with higher successional plants and climax species (large, mature trees).

Early successional plants tend to have high rates of photosynthesis and respiration, high rates of resource uptake, and high light requirements, whereas late successional plants often have opposite characteristics.

Early successional plants may be superior colonisers but they are inferior nitrogen competitors. They are fast-growing, shade-intolerant plant species that typically colonise degraded and low fertility soils. As soils change into those with higher organic matter and nutrient content they are colonised by slow-growing, shade-tolerant plant species.

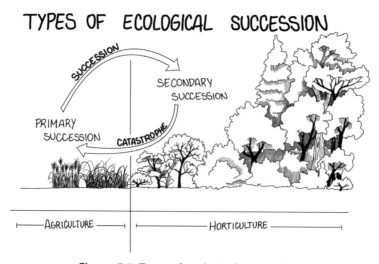

Figure 5.1 Types of ecological succession.

Early successional plant species invest in fine, longer roots for a more efficient nutrient uptake and in arbuscular mycorrhizal fungi (a scavenging strategy), whereas late successional plant species invest in microorganisms in the rhizosphere (a mining strategy) for nutrient acquisition.

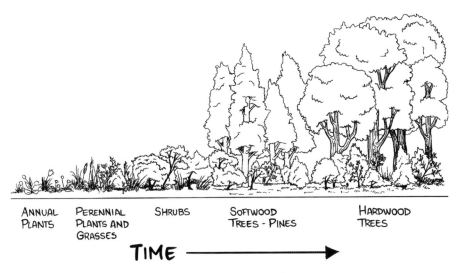

ANNUAL   PERENNIAL   SHRUBS   SOFTWOOD   HARDWOOD
PLANTS   PLANTS AND   TREES - PINES   TREES
              GRASSES

TIME ⟶

Figure 5.2 Succession is the slow progressive change in species over time.

In some cases, mycorrhizal fungi respond to late successional plants rather than early successional plants. However, the opposite can also be true in other cases, as it depends on what host plants are present rather than any abiotic factors and soil chemistry affecting the environment. Nitrogen-fixation is consistently different: It occurs at a greater rate in the early successional plants, and it appears that nitrogen immobilisation occurs with later (higher) successional grasses. Early succession favours bacterial soils, later mature forest favours fungi. This is discussed in more detail in Chapter 3 on soils, p.27.

Provided there is enough rain or irrigation, succession drives a landscape towards forest. Landscapes tend to be patchy, as disturbances (fire, storms, even clearing) always occur and interrupt the successional process. New bare ground becomes covered by pioneers (often weeds).

The whole significance of this is that when we garden and farm – plant our vegetables, trees, flowers or other (useful) plants – we are essentially planting pioneers and creating a patchwork of mini-ecosystems all at different stages.

In essence, ecologically speaking, our farms and backyards, which we enthusiastically tend and maintain, just want to grow up. We strive to create a mature woodland or open forest, or even a rainforest in that climatic region, but what we end up doing is more akin to a very young ecosystem that is never allowed to fully develop.

In the same way, we can think of sustainability as the delicate balance of negative and positive: A balance between things that degrade and damage to those that improve and enhance. Sustainability, then, is the equilibrium where systems are maintained, some damage is repaired and some good soils degraded, but may still be productive. Every time we dig the soil to harvest potatoes or carrots we cause harm, but unless we do this we don't eat or cannot make a living.

The Plant Layers of a Forest

There is some debate about how many layers or storeys of plants exist in a mature forest. A forest may have seven different layers of plants – canopy, understorey, shrub, herbaceous, root, ground cover and climbing. However, in some forests, there is an additional component – epiphytes, such as ferns and moss which exist on branches in trees, and some scientists include extra layers such as aquatic/wetland and fungal/mycelial as well as the emergent layer – those trees whose crowns emerge above the canopy.

A brief description of each of these seven basic layers follows:

1. Canopy layer (Tall tree layer)

The canopy layer contains the tallest, mature trees (climax species) and may reach 10 metres (30 feet) or more in height. It can include tall fruit, nut and timber species, and the canopy of branches and leaves create a shaded area underneath.

2. Understorey layer (Low tree layer)

This layer is just below the canopy (some call it the sub-canopy) and often contains small trees 3-10 metres (10-30 feet), such as most fruit trees and cash crops like coffee, cotton and sugarcane. Nitrogen-fixing trees are useful additions.

3. Shrub layer

Usually only up to 3 metres (10 feet) high, this fast-growing, thick, woody vegetation contains small fruiting bushes, such as blueberries and cape gooseberries, as well as other flowering and useful plants.

4. Herbaceous (Herb) layer

Dominated by herbaceous (soft stemmed) plants, such as ferns, wildflowers, companion plants and medicinal and culinary herbs, all of which may need to be shade-tolerant.

5. Groundcover layer

This layer contains the forest floor and besides a number of very low plants it contains decaying leaves, fallen branches, and mosses and liverworts. Plants grown here are often shade-tolerant and grow closer to the ground in dense formations.

6. Climber/vine layer

Vines and climbing plants have special structural features for support, allowing them to climb and find light. Light tends to be limited in very dense forests and plants have many strategies to seek light for photosynthesis.

7. Underground (Root) layer

The underground layer is also called the root layer. Here, the area around the roots (the rhizosphere) is where soil biota live and work, decomposing dead plant and animal matter into various organic substances and releasing nutrients. This soil layer is crucial for the water and mineral uptake by plants.

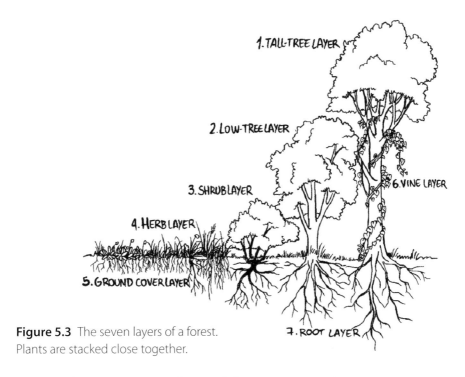

Figure 5.3 The seven layers of a forest.
Plants are stacked close together.

These layers exist to exploit sunlight, nutrients and water requirements. We can set up our farm spaces and gardens in much the same way – pack them close, provided we recognise their characteristics and requirements. For example, we can plant long-lived climax species (oaks, pecans, walnuts) with fast growing leguminous plants (acacia, albizia, tagasaste) and short-lived fruit trees (peaches, plums), with short-lived perennials (comfrey, yarrow) to provide weed control and shrubs (berries) and even annuals (pumpkins, beans).

The dense packing of plants in these layers is called stacking. When we plant we should consider planting reasonably close to one another in an area so that they support each other and provide ecosystem services. Stacking enables us to use space more effectively. Dense planted systems also reduce weed growth and soil erosion.

Masanobu Fukuoka (Japan) developed time stacking. Here, he starts the next crop before the last crop has finished. If we planted young fruit trees, nitrogen-fixing pioneers, windbreak, herb and vegetable species all together then we could harvest food for many years before the large perennials shaded out the annuals.

During this time the soil would be built-up due to years of mulch, compost production and green manuring.

Fukuoka proposed that weeds play their part in building soil fertility, so they should be controlled but not eliminated. He advocated using ground covers to shade out weeds or occasional flooding. There was no need for chemical control if natural balances were maintained. Growing healthy, sturdy crops in healthy soil was the secret.

Guilds and Guild Planting

A guild is a collection or grouping of plants, or plants and animals, that benefit when they are together. It usually involves assembling species around a central element – either a plant or an animal. So, guild plant or animal species are strategically selected to boost the productivity of that central element. The guild members, then, must either improve its yield or reduce the work needed to manage it. By design, this enhances the overall self-sufficiency and sustainability of the system.

For example, the main element might be a fruit tree and other elements are placed around it to help the fruit tree grow and develop. It is more than just companion planting, and also involves the concepts of interplanting, catch crops and rotations. (A catch crop is any quick-growing crop sown between seasons of regular planting to make use of the temporary idleness of the soil, to catch nutrients passing through or to compensate for the failure of a main crop.)

Advantages of Guilds

- Gives maximum productivity. e.g. carrots interplanted between onions. Onions are shallow rooted, tall-leaved and deter carrot fly, while carrots are deep rooted, have feathery leaves and deter onion fly.

- Reduces pest attack. Crops are disguised as strong-smelling neighbours to confuse pests. Tagetes marigolds fumigate soil of nematodes and umbelliferous plants (carrots, dill) host insect predators (wasps, hoverfly). Many predators of garden pests (e.g. wasps) only hunt to feed their offspring, themselves being wholly nectar feeders. Providing forage for adult stages is thus part of companion planting for a bug free garden (most small flowered plants provide this, including umbelliferous plants such as carrots, parsnip, fennel, dill and coriander, and others like various daisies, acacias and tamarisk).

- Insect eating birds (e.g. honey-eaters) can be encouraged by planting a few nectar producing and insect hosting plants (e.g. buddleia, banksias, dryandras, fuchsias, callistemon, salvia) scattered around your orchard and vegetable growing zones.

- Eagles and other birds of prey can be kept around to deter parrots and other fruit spoilers by keeping rabbits, pigeons or guinea pigs in your orchard. Alternatively, hawk kites flown overhead can be even more effective if not overused. A single alpaca or donkey amongst your sheep will keep foxes away. Most duck breeds (not Muscovy) will clean up slugs and snails and can be ranged through your food producing areas periodically when their appetite for seedlings will not compromise your yield.

- Fast growing plants can 'nurse' slower crops. e.g. tagasaste trees protect macadamia when it is establishing. Hardy nitrogen-fixing nurse species (e.g. honey locust, acacia, albizia, tagasaste) interplanted with orchard trees can moderate frost effects, improve soils, and provide mulch and shading for sensitive fruit trees such as avocado and citrus.

- Others planted as a windbreak bordering orchards (e.g. cane grasses, poplar, *Casuarina*) can be used to deflect or diminish frost, sunburn and drying or damaging winds.

- Allelopathy. Roots of plants can give off substances that can either hinder growth of other plants nearby or help them to grow and survive. As you gather observations, you might also come to notice that healthy apple trees are never found near walnut trees (walnut roots secrete growth inhibitors that apple trees are sensitive to).

- Providing variety – for diet and nutrition or for sale. It's an interesting fact that plants that make good companions often taste great together too! So growing them together not only improves their yield but also simplifies the job of harvesting. (e.g. marigolds grown with tomatoes, parsley and basil deter nematodes and contribute petals to eat in salads; dill grown under apple trees host predatory wasps and tastes great with apples raw or cooked).

- Reducing root competition. Grasses hinder fruit tree development, but fruit trees thrive in most herbal ground covers.

- Provide nutrients to fruiting crops. e.g. legumes provide nitrogen, some deep-rooted herbs mine potassium, phosphorus and calcium from the soil. Plants can also be slashed as a mulch (chop and drop). Plants that either act as a living mulch (e.g. nasturtium, sweet potato,) or shed mulch onto the soil (e.g. Bana grass, poplar, bananas) to form a protective cover, thereby improving soil conditions and retaining moisture.

- Foraging animals periodically allowed into the system also provide nutrients in the form of manure.

- Ask yourself 'How did grasslands work so well for millennia, feeding herbivores without the use of fertilisers and pesticides?' Pastures should function like grasslands and animals are part of that system.

- Remove pest habitat. Larval forms of orchard pests such as fruit fly flourish and multiply in fallen fruit, so seasonally introducing a forager such as pigs or poultry aids in pest control while adding fertiliser to the soil.

- Facilitating root penetration. Unlike grasses, some plants, such as comfrey, winter and spring bulbs, and globe artichoke, offer an open root structure that does not interfere with the central plant's ability to feed at the soil surface. Such plants should be established in orchards in place of grass to boost productivity.

Principles of Guild Design

1. Timing. For backyard gardens, work on a half-year rotation, so that harvest, weeding, fertilising and planting occur biannually. For larger projects plant in mild weather, preferably in winter or the rainy season so that plants have time to establish and there is time to set up the irrigation as required.

2. Avoiding bare ground. Fast-growing plants shade and protect the ground, and then are superseded by slower-growing species, e.g. lettuces mature before tomatoes, radishes before potatoes, nitrogen-fixing shrubs and groundcovers before larger climax species.

3. Companion planting. In the backyard garden, tomatoes, lettuces, cabbages, beets, carrots and parsley are all good companions (as are tomatoes and basil). For larger projects, use herbs under fruit trees, and plant nitrogen-fixing trees and shrubs to act as nurse trees alongside the slower-growing fruit and nut trees.

4. Rotating heavy and lighter feeders. Heavy feeders (e.g. leafy crops) are preceded by chickens or legumes and followed by lighter feeders such as fruit or root crops. When different stock animals are available it might mean cattle first, then sheep come in and clean up, and then poultry.

5. Mixing species. For backyard gardens, don't put plants of the same family together (or after each other) – spread them out. e.g. separate broccoli from cabbage and Brussels sprouts. For rural properties, plant a large variety of different shrubs and trees to provide a varied diet for stock or to ensure protection of the climax species of plants as they become established.

6. Minimising soil compaction. For backyard gardens plant plucking greens close to a path, then plants that can be harvested over a few weeks (tomatoes, capsicums), and then those plants that are harvested only once (broccoli, potatoes, beetroot) planted further in the middle of a garden bed. For rural properties, move stock through a fodder area quickly and minimise machinery operation in tree belts and so on once they are planted.

7. Planting small quantities of a large variety. Avoid glut or shortage by planting only a few of many different vegetable/fruit crops so that harvesting occurs over a long time, providing variety in our diet at all times. This is applicable in any farming operation.

Types of Associations

Interactions can either be positive (+), have no effect (o) or have a negative influence (-). A matrix of interactions is shown in the table below. As an example, in an orchard apples and walnuts might have a '-o' relationship – apples suffer. Apple next to mulberries produces '+o'. Mulberry next to walnut produces 'oo'. So, a sequence would be apple, mulberry, walnut (+oo). Now acacia next to walnut is 'o+' and acacia next to mulberry is 'o+'. Thus, the planting of apple, mulberry, acacia, walnut gives us '++o+' – the whole guild assembly seems to benefit.

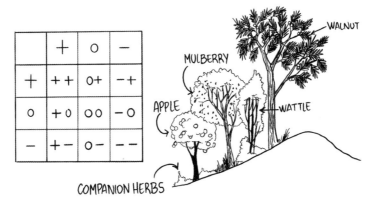

Figure 5.4 An example of a plant guild.

How do we Know what will Benefit what?

Companion planting guides and other references offer a great starting point to beneficial guild assemblies for any landscape design.

Observation is the only way to build upon this knowledge. You might even conduct a survey of plant and animal associations in your local area to this end. Keep a lookout for 'accidental' guilds that you can emulate by design: You may notice, for example, that a neglected but flourishing apple tree is growing alongside acacia and mulberry, with comfrey, nasturtium, iris and clover beneath it.

Interactions, both positive and negative, may or may not be sensitive to the distance between elements. Again, observation will provide the answer of how critical spacing is in your landscape design.

Little data is available on successful guilds so a lot of planting will be experimental. Hopefully, published knowledge will become available to all so that everyone can have successful operations.

Forage or Fodder?

Animals, like humans, cannot live on one food type alone, they need a varied diet. Plants that feed livestock can be classified as either forage or fodder. Forage is any plants that the animals graze themselves in a paddock, such as the various grasses, cereals and other shrubs that either naturally grow there or are sown by the farmer to feed the stock.

Many farming operations only use forage to maintain sheep, goats, cows and horses, for example, and these crops also contribute to land conservation and rehabilitation, protecting the soil and reducing erosion and loss of topsoil.

The challenge some farmers face is how do we find grasses that can survive on very low rainfall (typically long summers), knowing that there is increasing resource efficiency when we increase grass production, and that results in a corresponding decrease in carbon loss and emissions.

Furthermore, it is very difficult to have a well functioning forage system for stock if you solely rely on annual plants. No amount of annual cover crop will be sufficient, you need perennials to be profitable. The growing of multispecies plants as pasture is not new, but we are borrowing old techniques and using them with new technologies.

Fodder, on the other hand, is what is grown and deliberately fed to animals. This enables the farmer to regulate and manage the animal's food intake. Many small animals such as pigs and poultry are fed fodder. The fodder can be grown in an area, typically fenced, and the animals brought in to eat and are then removed; or the farmer brings cut branches and vegetative material and places this in the pen for animals to eat. Both techniques are common practice.

While fodder and forage crops serve the same purpose, they might thrive in different soils, soil moisture and local environments and require different levels of care and maintenance. These crops are an important tool for farmers, provided the right crops are selected and carefully

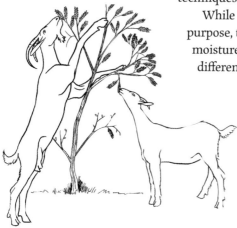

Figure 5.5 A variety of fodder shrubs and trees will provide high nutrition to animals.

managed during establishment and with grazing pressure to ensure optimum productivity is achieved. Forage might be more suitable for large farms while fodder is better for smaller farms that want to closely monitor animal food consumption. However, fodder lock-up areas with larger perennial shrubs and trees are ideal ways to feed larger stock when pasture is lean or becoming depleted.

Fodder crops can be either temporary or permanent crops. Temporary crops may be seasonal with multiple harvests, typically using a green feed of grasses, cereals and root crops as well as hay and silage. Permanent crops tend to be perennial plants which can be severely pruned and then allowed to reshoot. Not all perennials can take continuous cutting so you may need to experiment a bit to get the right mix of plants.

Fodder from trees and shrubs is especially useful during drier periods and the autumn (fall) and cold winter feed gap when nutritious food is not always available. Fodder plants should not be seen as just providing food during shortfalls, but also as positively contributing to an animal's growth and weight gain. Ideally fodder plants should be:

- Long-lived, fast-growing plants

- Reasonably drought tolerant (possess extensive root systems)

- Waterlogging and salt tolerant

- Able to be severely pruned or coppiced and then recover

- Highly productive, nutritious, digestible and palatable (pleasant to eat)

- Cheap and easy to establish, grow and harvest

- Multifunctional (fire retardant, N-fixing, windbreak, useful shade, timber)

- Providing leaves, pods or seeds in summer and autumn when stockfeed may be limited

- Not toxic nor inhibit nearby plants and pastures (allelopathic relationships).

Unfortunately, many fodder species are short-lived and do not have all of these characteristics, so it is difficult to find the perfect plants. Also, a particular forage may have low palatability in the wet season but become desirable in the dry period (or vice versa). As grasses dry out, digestibility decreases. This is because dry grass is high in cellulose and lignin (fibre) and low in carbohydrate and protein. Stock can then move onto other fodder species.

Again, don't rely on one main fodder species but a combination of several different trees and shrubs, along with pasture, is more likely to provide a good balance of protein, energy, carbohydrates and good nutrition. Often, mixtures of fodder plants have production increases of about 1 t/ha/yr for every 100mm increase in rainfall. What some farmers call a salad bar, a suite of forage plants such as a grass mix with chicory, plantain, red and white clovers, rye and timothy, does yield a high productivity rate. Some fodder plants, such as purslane, chicory, plantain, willow and oak also have natural anti-parasitic properties. So there are many factors and intricacies between plants and their consumers.

Wouldn't it also be great if we could find plants that reduced wind and soil erosion, required minimal water use to maintain their growth, improved soil structure, increased the biodiversity of the farm and did not themselves become environmental weeds?

Some characteristics of common fodder plants are shown in the table that follows. It lists examples of plants, but farmers will need to find plants that can be sourced in their area, are suitable as fodder and do not cause environmental issues.

Table 5.1 Generic fodder plants

Name	Region, rainfall	Growth, habit	Nutritional value	Other notes, uses
Carob – *Ceratonia siliqua*	D	Slow-growing, long-lived, spreading, evergreen tree. N-fixing. Male and female plants. Pods in autumn, 20-100kg/ tree.	Pods 60% digestibility, high sugar content (5% protein, 3% fat but 70% carbohydrate). Leaves palatable.	Pods ground to make carob powder, a substitute for chocolate. Good windbreak. Fire resistant.
Saltbush – *Atriplex* spp. e.g. *A. nummularia*	D	Spreading shrub. Most species 1-2m.	Some *Atriplex* species have poor digestibility and leaves contain too much salt. *Atriplex nummularia* 20% protein, digestibility 70%, up to 5 t/ha/yr.	Used in many arid countries as fodder supplement during summer and autumn.
Wattles – *Acacia* spp. e.g. *Acacia aneura*	D, C	Evergreen shrubs and small trees. N-fixing.	*A. aneura* 10% protein, moderate palatability and 50% digestibility.	Endemic. Acacias are also browse plants in Africa. Not all wattles are appropriate fodder species.
Oaks – *Quercus* spp.	C	Slow-growing, deciduous large trees.	Acorns 5-10% protein. 10kg/tree. 40-50% digestibility (contains tannins). Leaves also eaten.	Excellent timber, summer shade. Acorns drop in autumn when stock need extra carbohydrates.
Poplars (Cottonwoods) – *Populus* spp.	C	Some tolerate saline soils (Euphrates poplar). Tends to sucker. Less vigorous than willow.	White poplar *P. alba* – protein 14%, digestibility 77%. Up to 10 t/ha/yr. Some cottonwoods have 60% digestibility.	Groups include aspens, cottonwoods, balsams and subtropicals. Lightweight timber, windbreak, biomass for fuel.

Name	Region, rainfall	Growth, habit	Nutritional value	Other notes, uses
Tagasaste – *Chamaecytisus palmensis* (also *C. proliferus*)	C	Fast-growing, evergreen tree. N-fixing. Requires 500mm to be productive. Up to 10 t/ha/yr.	Leaves 15-20% protein. Highly digestible (70%) and highly palatable.	Seeds for chickens. Not tolerant of salinity, frost or waterlogging. Winter flowers for bees.
Tree medic – *Medicago arborea*	C	Fast-growing, evergreen shrub. N-fixing.	Leaves 15% protein, 50% digestibility. Available forage 0.5kg/plant.	Attracts bees and butterflies. Also suitable in semi-arid areas.
Willows – *Salix* spp.		Prefers moist habitats, tolerates heavier soils.	Up to 200kg/tree. Leaves 17% protein.	Fast recovery after harvest.
Honey locust – *Gleditsia triacanthos*	C, H	Fast growing, deciduous tree. Thornless varieties available, but seedlings likely to develop thorns. Tends to sucker.	Pods 10-100kg/tree. 17% protein, 60% carbohydrates (high sugar) and 7% fat. Pods drop in autumn.	Leaves are also palatable.
Indian siris – *Albizia lebbeck*	H	Deciduous, n-fixing Asian and African legume tree that tolerates wide range of soils and climates.	5 t/ha/yr. 20% protein in leaves, 60% digestibility.	Hardwood for furniture, useful shade tree, erosion control, bee forage (honey).
Leucaena – *Leucaena leucocephela*	C, H	N-fixing coloniser of disturbed ground. Can be weedy in some areas. Faster growth in higher rainfall areas.	1-4 t/ha/yr. Pods 25% protein, digestibility 80%. Leaves 20% protein, 75% digestibility.	Cattle forage in rangelands (but not as sole feed). Recovers from heavy browsing.
Pigeon pea – *Cajanus cajan*	C, H	Tends to be a short-lived perennial shrub to 3m.	Leaves 15% protein, 60% digestibility. Seeds 20% protein, 90% digestibility.	Overgrazing will kill plants. Best to prune to feed animals. Typically grown for human food with surplus feed to stock.
Elephant grass – *Pennisetum purpureum*	C, H	Tall, clumping grass to 3m. Bana grass (a taller *P. purpureum* hybrid) is also excellent fodder.	10% protein, young leaves very palatable, 70% digestibility.	Spread or propagated by stem cuttings or sections of rhizomes. Useful as windbreak. Kikuyu grass (*P. clandestinum*) is a common forage pasture.
Lablab – *Lablab purpureus*	C, H	Long growing season, tolerant of frost and very drought tolerant once established.	17% protein, digestibility 53%, 4.7 tonnes of dry matter/ha, shows weight gain of 0.6 to 0.8kg/head steer/day when grazing.	Low susceptibility to root diseases. Can be planted to a depth of 10cm.

Adapted from *How to Permaculture your Life*. Key: D = dry climate, rainfall typically less than 600mm (2ft), C = cool climate, rainfall between 600-1000mm (3.5ft) and H = humid climates with rainfall over 900mm (3ft) in subtropical and tropical regions.

Cash Crops

Cash crops are those that are grown for sale. In large-scale operations they typically include cereals like wheat, barley and oats or other crops like corn and maize, sorghum and soybeans, coffee, tea, sugarcane, vegetables, fruits and nuts, and non-foods like cork oak, tobacco, timber, rubber and cotton. Legumes are the second most important source of human food and animal forage, providing grains enriched with proteins, so these should be obvious inclusions.

Many countries, especially those with little natural resources such as oil and minerals, turn to cash crops which they export. Developing countries rely on cash crops for foreign income, but the crops are often planted on marginal and degraded land year after year which, in turn, produces problems with decreasing soil quality. When the best agricultural land is used for cash crops, local farmers are forced to use marginal land to grow food for the local market, and again this leads to further degeneration of the soil and environment.

Cash crops are distinct from subsistence crops which are those that a farmer grows to feed their own livestock or their own family. In recent decades, there has been a shift in what farmers grow, and the majority of crops now grown worldwide are those grown for revenue, with export bringing in the most financial profit.

Compounding the uncertainty of expected income is the volatile commodity market where prices fluctuate depending on supply and demand. What a regenerative farmer wants to know is 'what crop can I grow that is reliable, in demand, has a high value and makes the whole exercise worthwhile?'. Many farmers are turning to growing cover crops instead to feed stock which they sell, and relying less on growing monoculture crops in the hope of making ends meet.

Some farmers hold the view that anything that is not a cash crop is a weed, but this is simplistic and demonstrates poor understanding of the roles of plants in an ecosystem. At times it is a bit hard to grow diverse cash crops (commodity crops) but we can grow very diverse cover crops as sources of weed control, forage and for fertility.

Cover Crops

A cover crop is often grown in between normal crop production years. A cover crop may be sown to prevent bare soil or to improve soil quality, or both. The cover crop can also be fodder or forage (deliberately grazed), or slashed (chop and drop to mulch the soil) to allow soil microbes to have a feed and thus increase soil organic matter content and improve soil health. Even green manuring, where plants are grown and then turned (ploughed) into the soil, is helpful, although as we are now discovering that turning and disturbing the soil is detrimental.

Either way, killing or severe pruning of the cover crop plants and allowing their rapid decomposition is crucial to soil recovery and soil productivity. On the other hand, not every plant in the cover crop mix has to be edible as forage for stock. Plants have many ecosystem functions so we need some to attract pollinators, some to attract predators, some that are nitrogen-fixing and some to help mine nutrients from the soil. Ideally, we should allow pasture plants to flower and set seed because the plant will draw-up micronutrients to make these processes happen and there is thus more nutrition in seeds than leaves, which is beneficial to animals. And as we have discussed in a previous chapter, soil is the key to successful and profitable farming.

Cover crops can be warm season and cool season plants (cereals, brassicas), nitrogen-fixing herbs (chickpea, clovers, vetch), grasses (rye, millet), annuals (lupins, oats) or perennials (alfalfa, some clovers), and various combinations of these groups or categories. Many of these types of plants are known as forbs. Forbs are herbs that are herbaceous (not woody), broadleaf plants that are not grass-like. Rangeland or cropland forbs can be perennial or annual plants that are used to fill seasonal gaps in high quality forage. Examples of forbs include members of the brassica genus (kale, turnips, swedes), chicory (*Cichorium* spp.) and fodder beets (*Beta vulgaris*). Brassicas on their own are not advantageous as they do not have mycorrhizal fungi associations and they are not that good as stock feed either. Planting other seeds such as oats, rye and legumes with brassicas is a better option. Examples and characteristics of warm and cool season plants are found in the previous (Plant Function) chapter, p.71.

Ideally, several different plant species are grown as the cover crop. This gives variety to the stock diet, provides a range of mineralisation and soil enrichment processes and levels, meets the different needs of a wide range of soil biota, enables soil carbon to increase, increases water retention and reduces erosion and soil loss. When the soil mineral content increases then less additional fertiliser is required. This leads to saving on farm inputs.

Cover crops should be a mix of say six to eight species at a minimum (up to 10 is good) to provide the diversity to soil microorganisms and mycorrhizal and other fungi. Farmers just need to manage the cover crops so they don't get too high or too thick and dense.

While some regenerative agricultural farmers might have a dozen or more different species in the seed mix, this is probably not necessary, but at least consider the legume to grass ratio. For example, you wouldn't just plant rye, wheat, oats and barley but use one or two of these with peas, beans, vetch, and forage grasses. This is because some cover crops are nitrogen-hungry, such as cereal rye or triticale, and other plants can complement these.

Some other cover crops have long root systems to break up the hardpan. These include alfalfa, sunflower and chicory. Tillage radish is also known as daikon radish. This has specifically been bred for its thick, long taproot to help break up compacted soil. But being a brassica, it doesn't form mycorrhizal associations so it should be grown in a mix of seeds.

You can also change the species mix – the number of species and the relative amount of each – as a strategy to build diversity in the forage or soil conditioning operation. Generally, paddocks that are dysfunctional use lots of plants to build good soil structure.

What is clear is that increasing cropping diversity and using cover crops lead to significantly higher per-acre net profits, and you should always endeavour to use local, native grasses rather than expensive imported ones. This could include wallaby grass in Australia, switchgrass in America and Timothy in the UK and Europe.

There is no ultimate cover crop plant that everyone should grow. However, if you can find plants that have many or most of the following characteristics, and some are warm season and some are cool season, plant them:

- Germinates easily with little water/rainfall

- Fixes nitrogen

- Fast growing, broad-leaved to shade the soil

- Versatile, tolerates a wide range of soil types, pH and salinity

- Persists until it is no longer needed, easily killed or controlled

- Doesn't become a weedy pest itself

- Breaks down fast to release minerals back into the soil

- Low-growing so cereal or other cropping can occur at the same time (e.g. pasture cropping can be utilised)

- Palatable and nutritious for all types of stock

- Produces abundant seed which can be harvested or permit self-sown crops next year.

Whenever plants cover the soil, soil improvements are made: More carbon, more nutrition and greater cation exchange capacity, more soil structure and aggregate stability, better water holding capacity and water infiltration, more mechanical resistance to air and water erosion, and more biota can be supported.

We should treat cover crops like our cash crops. We should manage the cover crop to improve the soil to feed our cash crop. We need to keep the cover crop lower than the cash crop, so that the cash crop can flourish. This technique is the basis for pasture cropping, which is discussed in the next chapter (p.125). In essence, the cover crop is seriously mowed so it takes a long while to recover but allows the cash crop to grow well.

Even if your cover crop germinates but only grows 100mm (4in) and then dies or it falls over due to lack of rain, it still has roots which permeate the soil. The dead leafy plant still covers and protects the soil and while you may be frustrated that the plants did not reach their potential, nature will thank you.

The use of cover crops is increasingly becoming an integral part of cropping and farming systems in most regions in the world, although their adoption in arid and water-limited areas may put a strain on soil water availability for a cash crop. However, planting cover crops to eventually improve soils will undoubtedly increase water holding capacity and soil moisture. Good soil management and appropriate plant selection are the strategies to use in any regenerative agricultural endeavour.

Some cover crops (peas, buckwheat, sorghum and clover) are often seen as a waste of time and money, unless they can be harvested and sold or even turned into hay. If farmers can't get some monetary return they may be reluctant to sow. However, once soils improve and the inputs (costs) decrease then a small amount of money could be spent buying appropriate seed, and especially so if you run stock as these crops are possible fodder.

Farmers may be in a position to harvest seed from their cover crop and use this to re-sow or to sell. Very few seed suppliers provide mixed cover crop seed – you normally have to buy individual seed and mix it yourself. Often farmers would jump at the chance to buy these types of seed mixes.

If you are unsure about how much to sow, maybe start with a low seeding rate and see how this works on your farm and your soil. It is better to have a lower rate to manage the vegetation than to have a very dense cover crop which is getting out of hand.

A problem some farmers may face when buying seed is, where can they buy good seed without fungicides? What seed merchants provide worm juice coated seeds or use other biological seed treatments which are more beneficial for the soil? There are opportunities here for budding entrepreneurs.

Finally, there are four common farming practices that should never be practiced. Firstly, monocultures do not exist in nature, so growing monoculture single species is just plain wrong. Secondly, many farmers practice the concept of fallow, where a paddock is rested for a year or so. If the fallow paddock was sown with some cover crops or simply allowed to have weeds then this is far better than leaving the paddock bare. A paddock should never be left bare. The soil life

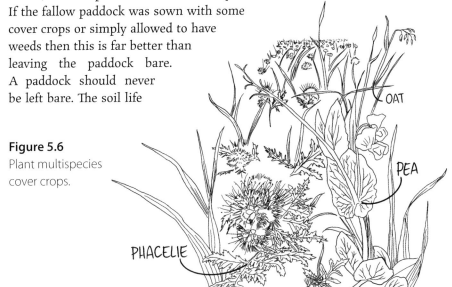

Figure 5.6
Plant multispecies cover crops.

needs to feed on something and a barren paddock devoid of plants will only cause the soil life to decrease.

Thirdly, don't plough as ploughing kills all types of plants and disrupts the functioning of the soil biota. Lastly, sometimes you should resist the temptation to hay a pasture as while stored hay is useful for times of drought in summer or in cold climates and snow in winter, cutting good pasture too soon could severely impact on soil cover, soil life and nutrient cycling.

Weeds

Weeds are contentious. They say that weeds are just plants growing in the wrong place at the wrong time or that we call a plant a weed if it has a detrimental economic impact on farm production.

Weeds are nature's way to cover and protect the soil, that's their job (or niche in ecological speak). Weeds flourish on disturbed ground. Weeds are pioneers and early colonisers of bare ground. We would like to think that nature allowed these weeds to change clay into loam or break up hardpan to improve drainage, but if they did we can't observe this in our lifetime. Succession can be a very slow and lengthy process.

In the context of regenerative agriculture, only think of a weed as a plant that your stock don't or won't eat, otherwise call some weeds forage. Is a plant that is in your field but is not part of your intended base forage really a weed? If it is causing yield or quality loss, then yes, it should be managed according to integrated pest management (IPM) principles. If it is a plant that contributes to forage volume or quality and can be grazed, then it contributes to your forage diversity and is not a weed. This is one reason why it is important to know the major plants in your pastures and their uses.

Agroforestry and Silvopasture

Agroforestry and silvopasture are similar, but with a twist. Agroforestry is essentially agriculture with trees. It is the melding of trees and shrubs into farming landscapes to enable the farmer to use trees to improve the environmental and financial values of the land. The trees and shrubs complement the crops grown in the 'pastures'. It's a marriage between forestry and agriculture, with the trees providing a range of services such as income (fruit, nuts, timber), improved soil health (nitrogen-fixing trees), crop protection (from winds, moisture loss), firewood, attracting predators and pollinators, and fodder.

While agroforestry is the generic term used when trees complement arable crops, trees around fence lines, salt scalds or rock outcrops are included. Alley cropping and intercropping are terms used to describe rows of trees (hedgerows) either side of the crop, so a sequence of alternating tree rows and arable crops is repeated in a paddock. Intercropping is a term some use to describe where multiple crops are grown together in the same basic area. In this case, you need to get the right balance of grassland and tree density.

- Silvopasture is agroforestry with the addition of animals, so in this case, silvopasture is the practice of integrating trees and shrubs with forage (such as cover crops in pasture) and grazing animals.

- Again, various trees and shrubs are integrated into pasture, cropping and animal farming systems to provide income from timber, act as windbreaks, provide food for stock and other services such as nitrogen-fixing.

- The trees and shrubs might be fenced off but animals are allowed in to prune fodder trees, or they can be just left to graze as they wish. Animals will under-prune the trees heavily so any trees will need to be well-established if you want them to survive.

- As you would expect, it is not as simple as throwing some animals into the paddock and planting some trees. Careful thought and design are needed to ensure the desired outcomes are met and the system becomes robust and resilient. In particular, you may like to consider these ideas:

- Silvopasture can be established in both existing woodland or bare paddocks. Any land is suitable provided the right trees are selected, although you wouldn't dabble with mature forests.

- Trees can be planted in rows, groups, or evenly spaced. Fast growing species such as locust, alder, willow and poplar offer an advantage because they can quickly grow above browse height, which allows for faster integration with grazing management.

- Place trees close enough together so that some of the pasture is shaded. This helps both the plants and the animals, which will assemble under the trees to avoid the sun.

- Animals should be matched to the land type and stage of plant succession. Animals can be destructive and can easily uproot newly established plants, trample young herbs and heavy animals can damage roots and compact soil. Animals should be brought into the system only after plants are established and/or trees are screened or fenced as required.

- Animals should be regularly moved in rotation so that after plants are grazed they have time to recover. This is crucial and the success of the silvopasture rests with animals being managed through rotational grazing.

- Trees should have multiple functions, products and/or uses. The permaculture principle asks you to think of, say, three functions or uses while I like to think of five. So planting mulberries provides shade, edible fruits, high protein feed for stock, windbreak and shelter, firewood from prunings and as a fair fire-retardant tree. Some oak trees provide nutritious acorns during the autumn feed gap, have high protein edible leaves, are drought-hardy, produce high quality timber, provide habitat, nest sites and food for a large number of insects and animals, and possess medicinal value for both humans and stock. Don't just think about trees as shade, windbreak or firewood, but choose them to provide as many functions, products and uses to the regenerative system as possible.

- Animal health and wellbeing is paramount. In some farming operations, stock only visit the feed bin for grain or hay. This severely limits their diet in terms of nutrition, and this behaviour restricts the animal's innate characteristics to walk, move and forage in the landscape. Animals need to be animals in the full meaning and exhibit the normal behaviours they would have in the wild.

- Create a complex ecology. While managing such a system has its own challenges, the benefits of providing a diverse, multifunctional woodland or forest amongst grazing animals far outweighs the time and resources needed to get such a system established.

- The system can be optimised by stacking inputs and outputs in both time and space, so there is always some feed all year round, there are always some trees providing shade (some are deciduous, some evergreen), there are trees and shrubs at different successional phases, there are always flowers for pollinators (so choose a number of plants that flower in different seasons) and there are some trees able to be removed for timber or for other reasons and new ones can then grow up and shine.

Finally, silvopasture should not be confused with silviculture. Silvi is Latin for forest, so silviculture is about growing, harvesting and managing tree production in a plantation or forest, mostly for timber or woodchip. Farmers may practice silviculture to some degree but silvopasture and agroforestry are so much more.

Figure 5.7 Sheep graze amongst honey locust.
Cows grazing cover crops along with black locust and walnut.

Figure 5.8 Alleycropping is a combination of crops and rows of trees.

Alleycropping

As farmers and farming change, more innovative strategies to maximise productivity and income will be trialled. Alleycropping is not new but is still in its infancy of acceptance and implementation.

Alleycropping is essentially planting trees in rows, often on the contour, and then having pasture or crops in between the rows. If trees are planted on a contour you might find curves in the tree belts, but these add microclimates and microhabitats to the system and therefore can be more beneficial than trees in straight lines. Rows of trees are spaced depending on the slope of the ground, but may range from fifty metres to several hundred metres.

Alleycropping has been shown to improve the production levels of vegetation between the tree rows. It seems trees act as shelterbelts and windbreaks, reducing wind strength, creating microclimates that could be pockets of warmth and moisture, reducing evaporation from the paddocks, providing nutrients to the soil from falling leaves and branches and protecting pasture plants with shade. Alley cropping has many other benefits too. It increases the biodiversity of cropland, creates new habitat for wildlife and provides shelter for stock during adverse weather.

The success of alleycropping depends on understanding how the trees and pasture or crops work together or compete for the same resources. While the amount of pasture or crops is reduced because of the tree rows, the productivity of that same area increases and there are many benefits to this planting system.

Selecting the right trees and the right crops is crucial, as extensive roots from trees may rob the nearby soil of nutrients. Ideally, the root patterns of the trees and crops should be different. For example, a grass pasture with shallow roots may work well with trees that have deep roots. In this case, they won't compete as much for water and nutrients. You could also choose plants that require water and/or nutrients at different times of the year. For example, there could be a cool season grass alongside a warm season tree, or vice versa.

Here is an opportunity to plant trees that could produce income, such as cabinet timbers, hardwoods or nut trees, and the selection of these should complement the speciality crops, cover crops or cash crops that are grown.

Pollinator Strips

Pollinator strips are areas on a farm that are planted with flowering plants that contribute to ecosystem services that build farm wealth. Ecosystem services and functions like water and nutrient storage capacity, carbon sequestration, pollination services, pest control, improving air and water quality, waste decomposition and protection against extreme weather events tend to be linked to the variety of species present and the multifunctionality of high biodiversity.

Many farmers are not aware of the need for other plants to complement the farming operation. Pollinator strips may be long lengths of vegetation along fence-lines and several metres wide, even outside of the farming property, that contain a large variety of angiosperms (flowering plants). Dense plantings ensure minimal ecosystem fragmentation and maximum availability of food and resources for wildlife.

Flowering plants in pollinator strips, which are not cash or cover crops nor used as forage or fodder, are important in all ecosystems and in the agricultural ecosystem they can:

- Provide food for insects, such as bees and butterflies, which are pollinators of many crops. If any crops have visible and colourful flowers then it is pollinated by an animal of some sort, and insects are on top of that list.

- Attract predators to the farm. Wasps, hoverflies and ladybugs prey on pests, so this helps to reduce insecticide use.

- Restore habitat for a variety of animals to increase ecosystem function.

- Filter wind and water. Pollinator strips are natural buffers and filter soil-laden wind and running water, keeping nutrients in the field, reducing wind damage to crops and cleaning water.

- Enhancing biodiversity. Biodiversity in any ecosystem ensures that the system will thrive and interactions between a wide variety of plants and animals can occur.

- Provide food for us. Without effective pollination of our crops much of the farm productivity would be lost.

Figure 5.9 Pollinator strips provide many ecosystem services.

The types of plants that are used in pollinator strips varies but farmers should endeavour to plant local, endemic flowering plants that are well-suited to that environment and can grow without too much care and maintenance. In many cases, we may need to deal with protecting our plants from adverse pH (acidity or alkalinity), high levels of toxic minerals such as aluminium, increasing salinity, and varying temperature and moisture levels in the soil.

Riparian Buffer

Riparian refers to the interface between a stream or river and the land alongside it. This zone is usually forested or planted out and thus protects the stream from the impact of adjacent land uses. Riparian vegetation along the banks of the waterway, also called streamlining, is a common conservation practice and helps with water quality, ecosystem enhancement and reducing pollution.

Riparian buffers intercept sediment and nutrient run-off from farmland, stabilise the banks of streams to minimise erosion during peak rainfall events and the various reeds, rushes and other wetland plants filter the flowing water.

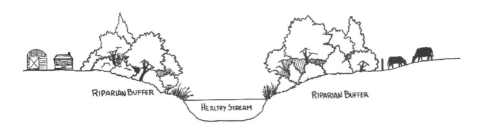

Figure 5.10 Riparian buffer strips can help keep the water in our streams clean.

The vegetation also provides shade, protection, food sources and habitat for a large variety of organisms which interact at this land-water edge.

Ultimately, a regenerative farm should work towards protecting aquatic ecosystems just as much as their own farmland.

A riparian buffer is usually made up of three zones, with the first zone at the water's edge and zones 2 and 3 progressively inland. Each zone may have a different width and mixture of plants, depending on the size of the water body and the desired functions of the buffer. Each zone of the riparian may protect freshwater and agricultural ecosystems in different ways. Zone 1 usually gets planted with large trees and shrubs, with reeds and rushes along the water's edge and even in the water. Some reeds can survive in full inundation, but others only in wet soil. Tree and shrub roots promote bank stability. The leaf canopy provides shade for much of the riparian area, and this shade can keep the water temperature cooler for fish and cooler for the animals that live nearby. Falling leaves and branches provide organic materials for both the stream and the understorey.

Zone 2 is typically a managed forest and/or shrub land, planted with native shrubs and small trees. Tree roots in this zone can slow down water flow from the adjoining land use, and this filters sediment and nutrients that might otherwise enter the stream.

Zone 3 is furthest away from the water source, right next to the agricultural land, urban area, parking lot, or even an industrial site. This zone is planted with grassland or a mix of grassland and wildflowers. It may be seen and act as a pollinator strip.

The design of a riparian buffer depends on a few factors, such as:

- adjacent land use

- slope of the land

- soil type

- current stream health

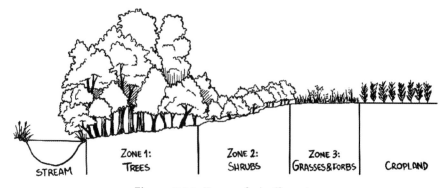

Figure 5.11 Zones of a buffer strip.

- the need for wildlife corridors
- width and size of stream or waterway
- availability of trees and shrubs suitable for that area and climate.

Once the required functions of the riparian buffer are known then it can be designed and built.

Wildlife Corridors

Wildlife corridors are vegetated connections across the landscape that link up areas of habitat. They allow the movement of species to find resources, such as food and water, and to find each other. With more and more clearing of land for farming and housing, natural areas of bush or forest become fragmented and many animals need a particular area in which to live. If that fragmented island area of remnant bush is small then animals might not live there.

Continuous strips of vegetation become corridors along which animals can roam. The corridors join lots of these fragmented areas, and animals have the opportunity to survive and thrive. They are the stepping stones across a landscape joining discontinuous pockets of forest or bush. Sometimes, pollinator strips, riparian buffers and alleycropping tree belts can also be wildlife corridors, enabling birds, small mammals, insects and reptiles to move from one area to another.

Like the vegetated areas just mentioned, corridors can also contribute to the resilience of the landscape in a changing climate and help to reduce future greenhouse gas emissions by storing carbon in both native and exotic vegetation.

To be effective, the wildlife corridor needs to be fairly wide, enough for animals to not be afraid to enter and travel through. Narrow tree belts are not sufficient. Somewhere between 20 and 50 metres is normally sufficient, but it depends on the main animals that wish to move so the wider the better. Wide, dense plantings reduce the edge effects of weed invasion and predator entry. Short corridors might be undertaken by individual landowners but large-scale corridors might span tens or hundreds of kilometres across multiple landscape types and jurisdictions. A large-scale corridor would require collaboration between a wide range of groups working in partnership to manage them.

Some management may be required if the corridor is built and plants need time to establish. The vegetation should contain plants typically found in ecosystems – some mixture of groundcovers, herbs, small shrubs, tall trees, vines and understory. These provide food, shelter and habitat resources, especially for endangered species. Like all vegetated areas, how do we maintain year-round green if there is no rain? We might have to irrigate, which may not be possible or economical, so there are challenges to face with keeping plants alive.

We can't all farm in high rainfall areas or places that receive some rain every month of the year, so we may need to find plants that are deep rooted and can

Figure 5.12 Wildlife corridors link remnant bush areas.

find enough water in the soil to survive in dry times or those that store water in their tissues (like succulents or cacti).

Wildlife corridors have an important role to play in maintaining biodiversity and ecosystem health, and although they only partly compensate for the overall habitat loss produced by the fragmentation of the natural countryside, they are part of the solution to regenerate the landscape.

Final Remarks

As a self-check on how well you are doing and if the paddocks are responding, compare your paddocks with what you observe along the road verges. Vegetation on road verges, albeit weeds and crop escapees, are often lush and thick. If your paddocks do not look the same then your management is well below what no management looks like (nature).

6

Regenerative Practices

Wherever you are in the world it should be clear that we need a lot of tools and skills to help us with the regenerative work on the soils we have. Farmland throughout the world is increasingly becoming subject to dryland salinity, soil acidification and loss of soil carbon. Farming is a complicated process and farmers need to manage their properties well, in order to make farming a viable business.

I think most people understand that many historical farming practices have had a negative impact on the environment, land and biodiversity. The destruction of entire ecosystems by human beings has been called ecocide, or murder of the environment.

Over the last century, conventional farming practices tended to be large monocultures of specific crops and often in association with particular animals, such as wheat and sheep. Another issue is that, like many other industrialised countries since the 1950s, many farmlands have increased in size while the number of farm owners has decreased, and more and more work is being done by ever-increasing use of larger machinery. In some regions of the world, crop farms are expanding and rangelands are decreasing. Ideally, we need the opposite to occur. In today's economic climate it makes sense to engage in lots of different enterprises to create business stability.

What we call modern agriculture really developed post World War II, and the current practices of chemical use, tillage and large-scale monocultures, with little regard to soil, erosion, waterlogging and salinity, has continued to grow.

A small number of farmers are doing things differently, and there has been a slow conversion of farmers to more holistic methods and practices based on organic farming, or to methods that build soil rather than denude the landscape.

While conventional cropping practices required that all weeds and vegetation be killed (usually by herbicides) before sowing and during crop growing, there are many farmers who now practice no-till, minimum till and a whole host of different techniques that build soil. Building soil, farming based on the principles of ecology and better farming practices, has become the basis for regenerative agriculture.

The obvious question to ask is 'why are there so many different approaches to regenerative agriculture?'. How come there are a lot of disciplines and practices that operate in this space? There are a lot of proven techniques and practices that do contribute to improved pastures and soil health but many of these are driven by individuals, so the movement is often fragmented and directed by personalities.

Complicating this is that many of these different practices are modified over time and they seem to meld together a bit. Certainly some farmers who are working towards a regenerative agricultural system pick and choose what they think suits them and their property, so it is rare to find anyone just doing pasture cropping or holistic management or natural sequence farming. And that is okay.

Cornerstone Practices

Historically, we often view farms as a series of paddocks with distinct areas for distinct purposes, but we should see the farm as a whole and all paddocks can be multifunctional in both grazing and cropping for a cash crop. Most of this new era in agriculture has only been developed in the last 50 or so years, and it hasn't been as a wave of innovation or milestones of innovation which implies a sequence. It hasn't really been a series of stepping stones where one innovative idea developed and then that was expanded and some new ideas added as time passed. Many of the practices were developed concurrently and independently. For example, while organic ideas have always been around, holistic management and permaculture mainly developed and took hold in the 1970s, pasture cropping in the 1990s while biological farming techniques are more recent.

However, there have been distinct pillars of innovation where certain practices have significantly contributed to the development of regenerative agriculture. What is common to many regenerative farming operations is most, ideally all, of the following six cornerstone ideologies and practices, with a quick comment here and elaboration to follow:

1. Organic. Chemical-free, natural amendments and fertilisers, biodynamics as an offshoot. Up until the advent of fertilisers, pesticides, herbicides and mechanised machinery, agriculture was undertaken organically with the use of animal manures as fertiliser, simple mechanical (physical) pest control and the occasional encouragement of biological control with beneficial organisms.

2. Regenerative grazing practices. Holistic Planned Grazing®, intentional stock movement, rational grazing (see page 116), strip and cell grazing, grazing naturally and other rotational grazing. Polyface Farm.

3. Holistic site design and planning frameworks. Permaculture: design of the whole site, integration of multifunctional elements, guilds and stacking. Holistic Management®, Regrarians Platform®.

4. Water capture and movement. Keyline, natural sequence farming, regenerative water harvesting and watershed restoration.

5. Regenerative cropping practices. Pasture cropping: Cover crops amongst a cash crop, multispecies forage crops.

6. Biological farming. Using biostimulants, earthworm juice, compost tea and biofertilisers, natural intelligence farming, MasHumus, biochar. Be biological farmers and not chemical farmers.

To see how each of these cornerstone practices have shaped the regenerative agriculture movement, let's examine each in turn, in no particular order. All of these practices contribute to the whole in their own way and are all equally important.

The discussion is not meant to be a thorough exposé but rather a short summary of the main ideas about the practice. Readers can investigate further if particular concepts are appealing.

Organic

What we eat, how it is produced and where it comes from can really make a difference. Essentially, organic growing and farming uses no synthetic chemical fertilisers and pesticides, and instead emphasises building up the soil with compost additions and animal and green manures, controlling pests naturally, rotating crops, and diversifying crops and livestock.

Organic gardens, pastures and orchards are both complex and holistic. It is not as simple as using compost instead of soluble nitrogen, phosphorus and potassium (NPK) fertiliser and garlic sprays instead of Malathion. A Rodale Institute long-term study showed organic farming produced the same quality and amount as conventional (chemical) farms but without the high money inputs.

The recent surge in interest in organic growing arose from the undisputed evidence linking the use of common garden and farm chemicals with human health risks and dangers to beneficial creatures in our environment. Organic is now mainstream and is recognised for its role in issues of environmental protection and repair, and the number of people (including farmers) converting to organic production is steadily increasing.

The basic principles to grow your own organic produce include:

- Building up the soil to be high in organic matter, nutrients and water-holding capacity.

- Controlling weeds and plants that harbour pests and disease.

- Rotating crops and using companion planting.

- Encouraging beneficial insects and other animals for pest control.

- The placement of plants is crucial (it's all about design!). Tall sweet corn might shade struggling broccoli, and you need to think about sun angles, shade and microclimate when positioning plants in the landscape.

The many benefits of organic growing include: Getting the nutrition you need, enjoying tastier food, fresher food, saving money, beautifying your community, fewer food miles, protecting future generations, protecting water quality, preventing soil erosion, saving energy (less transportation and pollution), promoting biodiversity, and using less chemicals (many herbicides, fungicides and insecticides are thought to be carcinogenic).

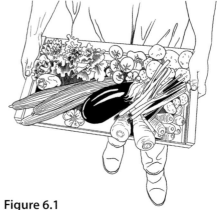

Figure 6.1
Organic produce fetches good prices in the marketplace.

Important differences are found between the biodiversity on organic and conventional farms, with generally substantially greater levels of both abundance and diversity of species on the organic farms, as the lack of herbicides and pesticides encourages wildlife.

If a farm is not being managed as an organic enterprise it is still industrial. This is because farmers still rely on agrichemicals to some degree. The industrial model doesn't take into account protection of bushland and forests, water and air quality, and conservation of soil.

Organic farmers might not use herbicides but they may till to control weeds and this destroys soil structure. Organic aims towards less harm, but various machines and procedures are used to control excessive weed growth. These include a roller-crimper and a row mower, and these are discussed in Chapter 8 (p.152) on integrated pest management.

Genetically Modified Organisms (GMO) contamination is another big issue for organic growers. Genetically modified crops have been known to spread into neighbouring organic farms and this causes severe problems due to contamination and the legal ramifications of seed saving, royalties and so on. The discussion here is not to debate the effect of GMO on food safety as the jury is still out and some GMO food products may impact human health. GMO itself may not be all doom and gloom and the GMO induced foodstuffs may be different. GMO foods might benefit humankind but the major thrust of the GMO industry is to enable companies to sell more herbicides and pesticides, mainly because the GMO enhanced plants are not affected but everything else is.

Finally, much of the cost of organic food may not be due to labour or other inputs but rather to simple supply and demand. There are not many organic farms but demand is high, and this demand outstrips supply in many places around the world.

Biodynamic Farming

Developed by Rudolph Steiner in the early twentieth century in Austria, bio-dynamics (also called anthroposophy) embraced agriculture and food production by considering spiritual, ethical and ecological philosophies. Essentially, bio-dynamics views annual crops and soil as a whole system, and it utilises various additives to activate and enrich soil, as well as the use of an astrological sowing and planting calendar.

Biodynamics is organic growing with a few twists. It relies on using particular sprays made from compost, minerals or herbs, or some combination of these, and planting at particular times based on the phases of the moon. Biodynamic farmers add nine specific preparations (numbered 500 to 508) to their soils, crops, and composts to enhance soil and crop quality and to stimulate the composting process.

Various herbs, such as yarrow, chamomile, stinging nettle, dandelion and valerian, are fermented or juiced as part of the preparation process. Cow manure is also used as one of the preparations, which is used as a soil stimulant before planting.

The second significant difference is the belief in the influence of the moon and planets to enhance germination, growth and production.

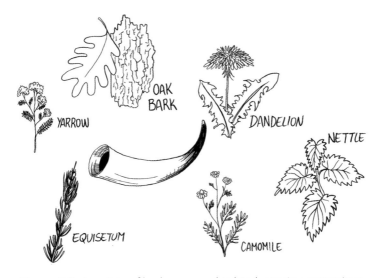

Figure 6.2 A variety of herbs are used in biodynamic preparations.

The effectiveness of biodynamics as an agricultural method is still unresolved, although many studies have shown improved soil life, greater numbers of earth-worms and plant nutrient uptake. However, generally, biodynamic farms produce lower crop yields than conventional farming, although the quality of food may be more important to some than more food.

Rational Grazing

Rational grazing was developed by Andre Voison in the 1950s and was the first method of grazing that was based on the idea of plants needing a recovery period after a graze. He used multiple paddocks per herd of animals and grazed them with multiple groups of stock following each other through the paddocks. He achieved very good results with animal performance and was able to still utilise paddocks intensively in the process. Voison was insistent that grazing needed to be intense to get a strong 'blaze of growth' as he called it. However, he emphasised the need for recovery periods following these intense grazes.

Rational grazing was based on rationing out the pastures via grazing slices and his set of four laws became the foundation of many other types of rotational grazing practices. These laws were:

1. The law of rest – grass must be given enough time to recover.

2. The law of occupation – the time that animals spend on a grazing slice must be short, so the pasture is only grazed once before stock are moved to a new paddock or grazing slice.

3. The law of maximum yields – stock must be offered the highest quality grass and pasture.

4. The law of regular yields – stock should only spend a few days, at most, grazing a particular paddock, and moving stock each day is ideal.

In a sense, Voison could be called the forefather of modern-day regenerative grazing practices where three main lines of practice have been developed as a result. The first is the practice of fixed recovery periods between grazing as is common with rotational grazing and some long rest period methods. The second, where graze periods are varied for fast and slow growth scenarios as is the practice in both Savory's Holistic Planned Grazing® and Parson's Cell Grazing. The third is the practice of varying recovery periods following grazing by deferring grazing or skipping some of the paddocks. This is the method that was preferred by Voison himself with his Rational Grazing Method.

Adaptations to rotational grazing techniques followed in the 1970s with Venter and Drewers in Southern Africa working on a five-paddock, five-year rotation with fire as part of that mix and more recently since 2000 by Dick Richardson with his Grazing Naturally Method of grazing which embraced a seven-year and seven-paddock or zone system, but without the use of fire as a management tool.

Rangeland and Pastoral Leases

Rangelands is a generic term that refers to areas of land which are most often uncultivated grasslands, woodlands and savannahs and are usually native herbaceous or shrubby vegetation. Occasionally, in some cases, other plants are introduced by humans. Rangelands are typically marginal agricultural land with low or erratic rainfall, as the better, more-productive land is used for agricultural crops.

Throughout the world, rangelands can also include wetlands and deserts, as well as the chaparral plant communities of western USA, the grassland steppes of Europe and Asia, the grass prairies of the Great Plains of central USA, and the tundras of alpine and arctic regions. In these areas the main difference between rangeland and pasture is management. Rangeland is simply managed by grazing whereas pasture is forage deliberately sown and cultivated as livestock fodder.

While rangeland plant communities are rich and complex, plant density can be low and vegetation sparse, so productivity is generally low and often these ecosystems are fragile. Generally, the carrying capacity of rangeland is typically low, so stocking density also needs to be low to enable native vegetation recovery.

A special mention must be made of a unique farming enterprise common in Australia and New Zealand. These are collectively called pastoral leases. About 44% of Australia and 10% of New Zealand is government-owned land (Crown land) which is leased to individuals or businesses. These areas are found in arid to semi-arid zones and tropical savannahs where conventional cropping and other more traditional farming operations would not succeed, mainly due to very low annual rainfall. This land has limited purpose and the main income focus is grazing. While pastoral leases, generally known as 'stations', are rangelands, the use and development of pastoral leases are not afforded the same rights as freehold land. As a comparison, a little less than 40% of farmland in the USA is also leased – not from the government, but from landowners who lease their property to others to use and farm.

Pastoral leases tend to have minimal inputs such as fertilisers and herbicides simply because of the cost for the scale of the operation and the requirement under the licence agreement to not plant anything that is not endemic. Pastoralists cannot plant, for example, cover crops and forage shrubs that are so common in other broadacre farming operations.

Pastoralists run stock (usually cattle, sheep or goats) which graze or browse on native vegetation. The lease agreement typically runs for a period of anywhere from 20-40 years to perpetuity, and stipulates that stock can only eat endemic species and pastoralists cannot introduce other fodder plants or forage crops.

The idea of land management can be contentious. We have to ask 'Are we wanting to manage a landscape to help the landscape or for our own benefit?'. Consider this: Set stocking is probably the most common system employed in pastoral leases and rangelands, but this can seriously alter the plant profile in the ecosystem. Palatable and highly edible forage quickly disappears, leaving plants that stock do not like to eat to thrive. While plants in these environments are tough enough to withstand harsh climates, many are long-lived but slow to grow. The necessity for good range management is heightened, as it takes a long time to recover if it's damaged or diversity is lost. With changing climates and a very wide spectrum of seasonal conditions, management systems need more flexibility and mechanisms to enable pastoralists to respond to maintain optimum vegetation cover.

Rotational grazing is essential to keep that plant balance, although in these large properties the fenced paddock might be hundreds or thousands of hectares (and acres). Moving stock might only be once every two or three months and this takes considerable time and effort just to round up all stock in that area and herd them to another paddock. What we have discussed already and is reiterated here is that plants need time to recover, and the longer the better.

Large numbers of edible endemic perennial and annual grasses and shrubs sustain stock. Some of these plants are eagerly sought after by stock and can be overgrazed, making recovery difficult if that area of land is not destocked.

Overgrazing ultimately affects water movement and the hydrology of soils, so managing stock by exclusion or rotational grazing is crucial for a successful operation. In particular, stock need drinking water every day and do not like travelling too far from their water source. Most pastoral properties are large enough that they are in a catchment of a particular river or waterway, or dams and bores are sunk to supply drinking water for stock. Having troughs or wind-mills in many locations helps stock to move over greater distances so that they do not eat out the most edible plants near the trough and increase grazing pressure on sensitive and vulnerable plants.

As rangeland can be fragile, there is additional grazing pressure from kanga-roos, bison, deer, feral goats and pigs, and rabbits – which are all uncontrolled grazers. It is interesting to see that in some areas of the world top order predators are encouraged to reduce these unwanted foragers. In Australia, some pastoral-ists are encouraging dingoes to control vermin, depending on what stock is kept. Dingoes don't usually attack cattle but can reduce sheep numbers so this unique solution may not work for everyone. In Europe, Asia and America, wolves, as apex predators, fill a similar niche by hunting hoofed mammals such as deer and bison, as well as hares.

What is clear is that both dingoes and wolves generally select the slowest or weakest prey, mainly the oldest animals or those in poor condition, although young animals are also taken. As top predators they keep the number of herbi-vores in check ensuring a balance in various animal populations. Most often graziers shoot dingoes (and no doubt wolves in some circumstances) with the misconception that they are wild dogs which is not the case. Ignorance and mis-information is no excuse for misguided action. Every day nature never ceases to amaze me. Whenever I travel through natural ecosystems I realise that we often underestimate, and certainly under-appreciate, the biodiversity and richness of species that exist. Even a sparse rangeland that seemingly only has a few types of shrubs and trees, holds a multitude of different grasses, groundcovers, various layers that you would expect in a woodland and an assortment of all types of birds and other animals, each playing an important role in food web interactions and providing a host of other ecosystem services.

The protection and rehabilitation of rangelands and pastoral lease stations is essential and can help reduce the impact of climate change by providing extensive carbon sinks as vegetation recovers and prospers. Furthermore, the incidence and frequency of desert storms, wind and water erosion events and the subsequent topsoil loss is reduced whenever good soil cover and stabilisation occurs. More innovative work is required to showcase best practice in the sustainable use of rangeland and pastoral leases, and the principles and practices of regenerative agriculture will be an important part of these strategies.

Holistic Management®

The idea behind Holistic Management® was based on Allan Savory's observations of herd animals in the 1960s, which moved throughout savannah grassland in southern Africa. He noticed that dense herds moved each day, pruning the grasslands, and then depositing tonnes of manure in that area. Birds moved in to pick through the insect grubs in the manure, thus breaking up the pads, tilling the soil and minimising pests.

All of this was essential for the health of the grasslands and he applied what wild herbivores did in nature to domestic livestock, recognising that changing

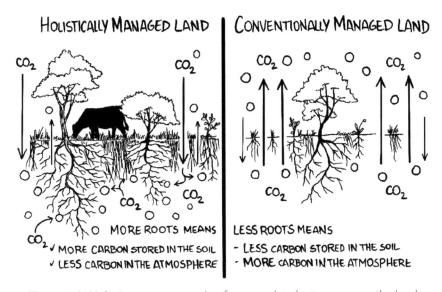

Figure 6.3 Holistic management is a framework to better manage the land.

one aspect of a system affected other parts of the system, and that each environment was unique. Subtle changes to the time herds were moved or areas allowed to recover were adopted, and it became clear that properly managed livestock was ultimately crucial to the health of the land.

Savory drew on Voison's work and recognised that time was more important than numbers, and it was not about the time you graze but rather the time given to plants and the soil to regenerate and recover after grazing. The secret to the possible effects of herbivores on pasture is management – when to move the animals, how long the grasses need to recover and so on. The movement of animals is, therefore, time-controlled, and adjustments can be made depending on the rate of plant growth, which varies from species to species.

Holistic Planned Grazing® is a management tool that fits within the Holistic Management® framework. It involves treating farm stock animals as a herd, bunched together and moved just like wild animals do when threatened by predators.

Grazing farm animals are herbivores. But it is not a simple matter to just let them roam about the paddock, as this leads to poor soil and plant management. Electric fencing enables farmers to control and move stock in much the same patterns in nature. There also needs some further considerations, such as adjusting the stocking rate to suit the growth rate of the forage and providing a diversity of plants to improve ecosystem services.

In nature, birds follow herbivores, and they mix and turn over the soil and thus stimulate vegetation. So the ideal for a farmer is to have a small group of herbivores followed a day or so later by a larger group of birds. On a farm, chickens are best for this.

Holistic Management® is a decision-making framework that considers the economical, social, and environmental impacts of any decisions farmers make. Holistic Management® uses a set of testing questions. These basically ask farmers if the actions they are taking will address the specific problem, if all family members are involved in the decisions-making process, will their actions decrease profits (or increase costs significantly) and do the actions improve the quality of life for every living thing in the farming enterprise?

Cell Grazing

Cell grazing is a variation of strip grazing and rotational grazing. Rotational grazing is moving stock from paddock to paddock, but the paddocks may still be continuously grazed – the animals are not confined to a very small area and they can wander over a larger paddock.

Strip grazing is a strategy employed mainly in winter where animals are confined to a 'strip' of paddock for a short period of time. It is a more intensive form of rotational grazing where the pasture is rationed to the animals, so a lush area of paddock can hold a higher stocking density than normal. If animals were allowed to roam then much of this luxurious growth would be trampled or spoiled by dung and thus be wasted.

Cell grazing can be thought of as a more intensive form of rotational grazing. Paddocks are divided into smaller areas (cells) and animals moved from one section to another. It has long been recognised, since Allan Savory's observations of herd animals in Africa and his work with Stan Parsons, that pastures need moisture, sunshine and rest to recover from grazing. Since farmers have no influence on rain or sun, at least they can control the 'rest' time the paddock needs to recover and regrow.

Figure 6.4 Electric fencing is portable and contains stock in tight bunches.

Some grass species need a couple of weeks to recover before they start to grow, so leaving paddocks untouched for three to four months is the minimum rest period. It is not a simple matter to rotate animals every three months or longer, as plant recovery and regrowth is often seasonal, temperature and rainfall dependent, and at times erratic. In some agricultural areas you may have to rest the paddock for a year or more. What is clear from Savory and others doing similar work on the ground is that every farm is unique and has to have its own management plan.

Animals should not be allowed to graze the plant material to the ground. There should still be sufficient leaf height above the ground so that plants can use sunlight to make the many compounds needed to produce new tissue.

It is far more energy efficient for this process to occur rather than wasting root reserves to enable plants to recover. So the overall thrust of cell grazing is high stocking rate in small areas for a short period of time (typically one to three days depending on the size of the cell).

Advanced Cell Grazing

Developed by Geoff Lawton in Australia, this adds other dimensions to moving and utilising stock. Here, a fenced laneway around the property is used to move stock to various 'cells' or paddocks, as well as incorporating swales and drains to collect manure washed away in run-off from rain events and direct this to plants in adjacent areas. This fertiliser then increases the growth and productivity of nearby pastures, forests or orchards.

Whereas cell grazing has traditionally been animals on grass pasture, advanced cell grazing can incorporate sections of forest, specific fodder areas, grass and any other combination of these. This gives animals the diversity (and nutritional variety) in their forage, and ultimately, you would think, better health seen in animals.

Polyface Farm

Polyface Farm is the property owned and developed by Joel Salatin and his family in rural Virginia, USA. Salatin extended Allan Savory's idea by using different animals in sequence. Cows initially fed on pasture for a day or so and then they were moved. Chickens were bought in a few days later to feed on grubs in the manure.

Polyface Farm is very animal-centric, with pastures being the main plant focus. Besides cows and chickens, pigs, rabbits and turkeys are used as part of the farm management process. Timber is also harvested from their forested areas. Some other philosophies, such as permaculture, promote a wider range of plants, which normally take up much more of the farm design. By stacking complimentary enterprises on the land base, such as what is achieved on Polyface Farm, you can increase farm profitability.

Figure 6.5 Chickens follow larger mammals in the paddocks.

Salatin describes the forage as a "fresh salad bar" as animals are allowed access to a concentrated area of fresh grass so that grazing occurs in a small area over a short period of time. Thus Polyface is more akin to stacking in time as the animals' rotations are scheduled. The overall principle is simple – using mobbing, mowing and moving, with long intervals in between times of very controlled disturbance.

Salatin also recognised that grain should not be fed to stock (herbivores don't naturally eat seed), and if this occurs it is an indication that the pasture is inadequate, deficient or inappropriate in some way. If we grew the right types of plants and fodder trees, there wouldn't be an 'autumn feed gap' as feed would be available throughout the year. Not all stock animals eat every type of fodder plant. Some animals are very particular about what they eat, others less so, and some animals have to become accustomed to eating something different.

Polyface lets animals do what they are good at – their natural abilities and habits can replace machinery. Pigs are great 'tractors', ploughing land and clearing weed infestations, chickens can remove both weed seeds and insect pest larvae from the soil, and earthworms aerate the soil as they burrow and produce amazing fertiliser as they digest organic matter. Pigs are arguably the ultimate disturbance tool.

Grass recovery is aided when secondary animals, such as chickens, scratch and perform other work to break-up and spread manure, eat weed seeds, parasites and insect pests, and aerate the soil.

With thoughtful management of pastures, supplementing perennial grasses with a range of fodder and medicinal trees, ensuring the right stocking density and providing enough water for the stock, there may not be a need to buy in additional feed such as grain or hay.

Polyface farming, along with cell grazing and strip grazing, is very labour intensive, whereas continuous grazing (called set stocking in some countries) uses minimum labour. Moving fences and stock every day, along with bringing in chickens or other animals makes a very long day for farmers who practice intensive grazing techniques.

Polyface has also pushed boundaries by marketing their organic produce to the local community. Their mantra is "beyond organic" and they sell as much as they can, and at times cannot meet demand.

Permaculture

All the aspects of regenerative agriculture can be experienced in a well-maintained thriving permaculture garden. Here you can see biodiversity, integrated systems, pollination, predators, soil cover and protection, companion planting, pioneers, food production, and a huge variety of plants that are used for screening, nursery trees, nitrogen-fixers, windbreak, predator attracting and so on.

Permaculture has various other principles and practices that enable holistic functional design to be provided. Not only is a design a landscape plan, but it may incorporate design strategies for both people care and earth care. It's about designing for resilience, regeneration and interdependence, but has the capacity for adapting planning and design as these can be fluid and can change over time.

The principles on which permaculture is based are drawn from the writings and teachings of the two co-originators Bill Mollison and David Holmgren.

The following list melds these principles, many of which are much the same but may be worded differently. They are shown side by side, with occasional differences as seen in the table.

Table 6.1 Principles of permaculture

Mollison	Holmgren
1. Relative Location – every element is placed in relationship to another so that they assist each other 2. Each element performs many functions 3. Each important function is supported by many elements	8. Integrate Rather Than Segregate
4. Efficient Energy Planning – for house and settlement (zones and sectors) Zone planning, Sector planning, Slope	7. Design From Patterns to Details
5. Using Biological Resources – Emphasis on the use of biological resources over fossil fuel resources	5. Use and Value Renewable Resources and Services
6. Energy Cycling – energy recycling on site (both fuel and human energy)	2. Catch and Store Energy 6. Produce No Waste
7. Small Scale Intensive Systems Plant stacking, Time stacking	3. Obtain a Yield
8. Accelerating Succession and Evolution – Using and accelerating natural plant succession to establish favourable sites and soils	9. Use Small and Slow Solutions
9. Diversity – Polyculture and diversity of beneficial species for a productive, interactive system. Guilds.	10. Use and Value Diversity
10. Edge Effect – Use of edge and natural patterns for best effect	11. Use Edges and Value the Marginal 7. Design From Patterns to Details
11. Attitudinal Principles Everything works both ways Permaculture is information and imagination intensive	12. Creatively Use and Respond to Change
	1. Observe and Interact 4. Apply Self Regulation and Accept Feedback
In addition to the above, Mollison provided these others: • Work with Nature, Rather than Against it • The Problem is the Solution • Make the Least Change for the Greatest Possible Effect • The Yield of a System is Theoretically Unlimited • Everything Gardens	

As Bill Mollison once said, "You can change a desert into rainforest or rainforest to desert. The choice is ours." It's our choice! Permaculture seems to attract people who are creating the future, and devotees are changing the world, one person at a time and one property at a time.

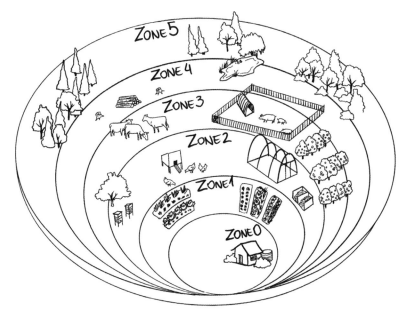

Figure 6.6 Permaculture places all components (elements) in particular zones.

Pasture Cropping

Based on the work of Coin Seis and Darryl Cluff, developed in 1993, pasture cropping advocates sowing cereal crops directly into perennial pastures and combines grazing with cropping. Pasture cropping is perennial cover cropping where a cash crop is sown into a dormant perennial grass. A multi-species crop that might include species such as radish, turnip, oats, vetch and field peas, can also be planted amongst endemic native grasses.

Paddocks are never ploughed, so that the native and exotic grasses can become established and stay that way. The aim is to have 100% ground cover, and this increases the soil carbon and biomass content and the soil life. Some devotees still use selective herbicides to control weeds and conventional fertilisers to change the soil chemistry.

When pasture cropping is first adopted, there can be a grain yield penalty until the soil ecosystem is restored. Production is often lower because some nutrients, such as nitrogen, are not cycling due to poor soil structure, and some nutrients are being utilised by the grasses.

This reduction in grain production is easily compensated by an increase in stock forage, which increases the profitability of that side of the farming enterprise, and restoration of the grassland or perennial pasture and soil ecosystem.

In some ways, pasture cropping is better suited for a grazier moving from a pasture system and not that suited for a farmer moving from a conventional cropping system.

Cereal crops can be sown into summer (C4) and winter (C3) perennial native grasses (see Chapter 4 (p.71) for a short explanation about C3 and C4 plants). When a winter cereal crop is grown in a summer perennial pasture, there are distinct growing periods for each species. In this case, and probably the majority of cases, C4 (summer) perennial grasses are used for the permanent ground cover. The cash crop seed is usually sown with a seed drill, cutting the sod rather than turning it.

Figure 6.7 Pasture cropping. A cash crop amongst cover crops.

Stock can graze right up to sowing and again after the crop is removed, and there is no need for a summer fallow. You could even sow a crop only every few years for both pasture and soil recovery.

Whereas conventional cropping methods sprayed all weeds and other plants all of the time, pasture cropping controlled weeds by selective use of herbicides, by livestock management during grazing, and the production of plant material litter to smother them.

It is essential to have complete ground cover at all times, and this results in large increases in plant biomass and soil carbon. However, if weeds are not reduced by some method, then production of both crop and grass is severely reduced.

No kill cropping, also known as advance sowing, is similar, but it does not eliminate weeds, preferring diversity. No herbicide, pesticide or fertiliser is used, allowing maximum biological activity and minimum input costs, whilst maintaining plant diversity.

Generally, productivity using no kill cropping is much lower than conventional farming but as the input costs are far lower, profit can be made. Developed by Bruce Maynard in 1996, no kill cropping relies on sowing seed in dry soil, minimising soil disturbance and managing grazing to support both plants and livestock, while inhibiting weed growth.

Keyline®

When rain falls there are only three options: It enters the ground, it runs off the land and enters streams and then rivers (and usually ends up in oceans), or it is allowed to be captured and stored.

Groundwater recharge is a great outcome as water is stored in the soil (although maybe a long way down), but having water leave the property and head downstream is not ideal. Run-off often causes erosion and loss of topsoil into waterways too.

The third option, the capture, storage and use of water, has always been another worthy consideration for fully designed properties.

PA (Percival Alfred) Yeomans, an Australian mining geologist and farmer, developed the keyline plan in the mid-1950s, with the aim to increase soil fertility, improve soil structure and to hold water in the landscape. He pioneered topsoil regeneration, on-farm irrigation dams, chisel ploughs, contour ripping and non-terraced flood irrigation. Dragging the digging points of the chisel plough through the soil enables water, air, seed and manures to penetrate the soil and provide the right materials and medium for soil life to proliferate. Yeomans also planted trees alongside the contours to better utilise the water that flowed there.

One of the typical benefits of keyline is the rapid development of living soil. Most of our fragile landscapes (what Allan Savory calls brittle landscapes) need hydration not draining. Engineers like making drains and moving water away as fast as they can, but ecologists like to keep water where it falls and hold it there, preventing ridges and hills from drying out too quickly.

The soil acts like a huge sponge and stores water within the soil matrix. Where there is water there is life, and the soil life emerges to begin all those elemental cycles to make nutrients available to plants and to all soil biota.

Yeomans identified a specific contour in a ridge-valley-ridge system where changes in the slope occurred and he called these the keyline, and the spot which is essentially the start or beginning of a valley, the keypoint. Once the keypoint and keyline are located and marked out, subsoil cultivation (minimum tillage) occurs parallel to this contour.

This forces run-off and shallow seepage water to collect in the cultivation channels and flow towards the ridges, thus keeping water higher in the landscape for a longer time (rather than letting water just drain down the slope towards a creek).

These cultivated channels above and below the keyline become off-the-contour drains, intercepting water as it moves downwards and redirecting it outwards towards ridges.

Yeomans also developed the Keyline Scale of Relative Permanence, which was a sequential list of factors that ranged from permanent in a landscape and couldn't be changed (climate, landforms) to those that could easily be changed on a farm (such as fences, planted trees and soil). This helped farmers plan where

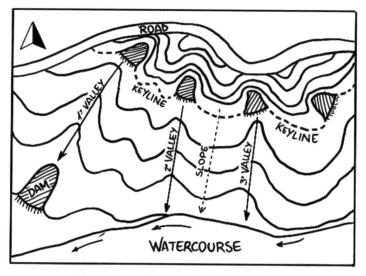

Figure 6.8 A keyline plan for a property.

to place dams, roads, buildings and tree crops. The scale of relative permanence, in decreasing order of permanence, is: 1. Climate; 2. Landforms; 3. Water supply; 4. Farm roads; 5. Trees; 6. Permanent buildings; 7. Subdivision fences; and 8. Soil.

The Yeomans Keyline Plow (plough) has fixed tines or shanks that slice through the subsoil and break up the hardpan below, all without turning the soil over like most modern ploughs.

Initially you might set the points in the tines at about 150mm below the surface and then progressively make deeper cuts each year and end up about 400mm below the surface.

Regrarians Platform®

Darren Doherty and Lisa Heenan have adapted Yeomans' scale of relative permanence into the Regrarians Platform® (from Regenerative Agrarian – an agricultural society). They have tweaked the eight factors in Yeomans' scale and added two more – economy and energy.

Whereas Yeomans' scale was centred on agriculture, Doherty and Heenan have included social and economic themes. This allows farmers to explore ways in which they can value-add to their products or seek new markets, and develop greater community resilience.

The Regrarians Platform® consists of these layers:

1. Climate – considers both the human 'climate' of those involved in holistic decision-making and the activities of the biospheric climate and the implications of that on farming enterprises.

2. Geography – using aerial and topographic maps, landform shapes and landscape capability in order to better understand the terrain and its limitations or advantages.

3. Water – makes life possible. The emphasis should be to allow water to do its duties, rehydrate land and reinvigorate the ecosystem.

4. Access – those arteries of movement and their engineering ranging from dirt tracks to major roads. Includes utilities such as electricity, telephone and water services.

5. Ecosystems – these vary from natural to created forms and their placement is defined by the previous four layers. Includes all plants such as those in crops and forestry enterprises, farm and other animals, and soil life.

6. Buildings – most often large infrastructures that require detailed planning and engineering, but also living spaces and portable infrastructure. Buildings are placed relative to the previous five layers.

7. Fencing – can be fixed or temporary to define paddocks, but with the advent of electric we can better manage landscapes by following topographic patterns and considering natural animal behaviours.

8. Soils – the focus is determining the most cost-effective treatments, along with testing and monitoring soils, to regenerate soil and this provides us with a strong foundation for our agriculture.

9. Economy – developing strategies to drive the economic engine of the various capital flows and to ensure our overall enterprises are viable and continue to be so.

10. Energy – the energy layer is not just about fuel for tractors and machinery, but all aspects of organisms and ecosystems which rely on energy (such as wind and solar) to exist: mainly through the role of plants to produce food, a range of substances and oxygen through photosynthesis.

These 10 layers of this agroecological landscape and enterprise planning platform are further categorised into:

- Constitutional and Foundation layers: Climate and geography

- Infrastructure and Development layers: Water, access, ecosystems, buildings, energy, fencing

- Management layers: Soil, economy.

Natural Sequence Farming

Peter Andrews (Australia) developed this ideology that mirrored what he saw in ancient landscapes. In some areas, a natural chain of ponds existed and seasonally the waterways and ponds overflowed and slowly spread across the landscape. This mechanism didn't allow the water flow to speed up and drain the landscape.

Flowing water incises the landscape and creates drains, whereas Andrews installed a series of leaking weirs to allow water to get through but more slowly so water backs up and spreads out over the landscape, causing it to slowly percolate through the soil and consequently improve soil structure and fertility.

Leaky weirs are made by putting rocks or logs across the creek to create a barrier to slow water down (not stop it like a dam wall), to spread it out and then allow water to slowly pass through downstream. The subsequent rehydration of the landscape (also called hydrolation) results in the natural vegetation to flourish.

While not necessarily organic in its practices, natural sequence farming recognises weeds as pioneer plants, which are allowed to proliferate and hold and build soil. After these are slashed to add fertility to the soil, palatable grasses become established.

Natural sequence farming can be practiced on one farm, but to be effective it has to be adopted by a catchment where all neighbouring farmers bring about change.

Regenerative Water Harvesting

A similar set of techniques, called regenerative water harvesting, induced meandering and along with watershed restoration, was originally developed by Americans David Rosgen and Bill Zeedyk, with later collaboration with Van Clothier, and more recently promoted by Craig Sponholtz (Watershed Artisans) and his colleagues in northern New Mexico.

They developed ways to reduce erosion and restore natural water movement in the landscape, by harvesting run-off in simple structures that allow water to spread throughout the soil, encourage plant diversity and thus build soil.

Craig's work focuses on using both modern river restoration techniques and ancient farming practices to restore (agricultural) land productivity by creating beneficial links between agricultural systems and natural ecosystems.

He uses rock mulch to slow water, prevent erosion and direct water into

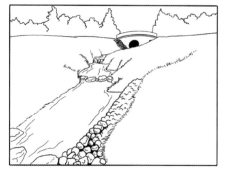

Figure 6.9 Watershed restoration manages water movement.

other areas, and then seeds a large variety of plants to continue with the soil restoration process. Increasing soil activity and soil organic matter makes noticeable improvements. For example, for every 1% increase in soil organic matter, the soil can hold an additional 100,000 litres of water per hectare (20,000 gal/acre) as the organic matter can hold about ten times its own weight. Regenerative water harvesting is more of an ecological approach whereas natural sequence farming is more of an engineering approach.

The guiding principles for bringing moisture back to land include:

- Find the places where moisture already exists to some extent and start there.

- Protect and expand the moisture storing areas of the landscape.

- Stabilise banks and slopes to prevent erosion and further degradation.

- Restore dispersive flow and increase soil infiltration at every opportunity.

- Cultivate restorative plant communities and build biologically active soils.

- Create site specific solutions using natural forms and processes. This might include harvesting run-off to keep it in the landscape before you attempt to move it, and to capturing sediment as a potential resource material.

- Once water restoration techniques and structures are in place, manage landscapes to increase ecosystem services and productivity.

Biological Products

When we grow crops and food, we often neglect plant nutrition. We forget that plants take up nutrients from the soil and the soil becomes depleted.

In all agricultural systems, the addition of fertiliser, in some form, to our soils is crucial to food crop production. We can make simple organic-based fertilisers from weeds, seaweed, manures, worm castings and various combinations of any or all of these. Biostimulants are covered in Chapter 3 (p.62) on soils, but here is some further information about what products can be used to provide easy solutions to nutrient-hungry plants and soils.

Basically, biological-based soil additives and plant foliar sprays are made by either fermentation or non-fermentation techniques. Let's have a look at some of these.

Fermented Fertilisers

The digestion of plant and animal material can be undertaken without air (anaerobic) or with air (aerobic).

Fermentation, by definition, is an anaerobic process, but we use it here to include any digestion of organic matter, with or without air.

In these preparations, bacteria and other microorganisms are used to break down organic matter into simpler substances that can be applied to plants and the soil.

Anaerobic Digestion

Some bacteria do not require oxygen to survive, and they are able to use plant and animal material as their food source.

When this occurs, various gases are produced as a by-product of the digestion process. These mainly include methane, carbon dioxide and nitrogen, but may also include sulphur oxides, hydrogen sulphide and ammonia.

All of these gases need to be vented from the system, otherwise gas pressure builds up and liquid and gas explosions can occur.

Manure is placed in a drum or container, water is added, along with some other ingredients such as molasses and milk, and allowed to ferment for at least one month and up to a few months depending on climatic conditions. Excess gases are passed out of the tank through a water trap, which excludes air from entering but allows metabolic gases to escape. If these gases cannot escape they build up in the system and inhibit microbe action.

Biofertiliser is a fertiliser that contains living organisms and is often made from manures. Most animal manures seem to work, including chicken manure or other bird manures with mixed success. Manure from grass-eating animals, such as cows, sheep and goats, seems better than that from meat-eaters, such as dogs. Pig manure is used to produce methane.

You basically produce a microbe culture to inoculate the soil and a solution containing readily-available minerals to feed the plants. You can 'seed' a culture with a particular type of microorganism (e.g. rhizobium, blue-green algae), or you can make a generic brew that contains microbes that are found in the manure, yeast, air and water that you add. You need to use the fertiliser well before the culture dies. Even so, adding minerals and nutrients to plants is beneficial every time.

To make a brew, add about 20kg (44lbs) of fresh manure to a 200 L (52.8 gallons) drum (with a lid that can be sealed). Fresh manure contains more beneficial bacteria and other microorganisms. Fill to about four-fifths with water.

Add small amounts of a range of other substances, including milk, yeast, rock dust and molasses, all of

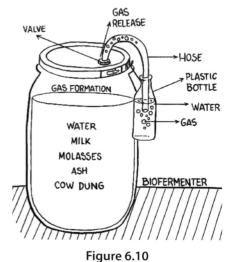

Figure 6.10
Anaerobic digestion makes biofertiliser.

which help to kick-start and feed the microbe population. Install a water trap and allow to ferment for a month or two (visible bubbling may stop after a few weeks or more). When you open the lid, beware! It may stink. If it does smell very badly then the product should not be used – it most likely contains pathogens. While biofertiliser does smell, it is not unpleasant – it should smell like a typical ferment.

Pass the liquid through a mesh screen or coarse cloth to remove the digested remains of the manure. The resulting liquid fertiliser can be diluted at least 10 times with water before spraying on plants and soil.

Latin American agronomists and agricultural engineers Jairo Restrepo Rivera and Sebastião Pinheiro pioneered training farmers in on-farm biofertiliser production and using round-filter chromatography to determine soil, crop or compost quality. MasHumus (meaning 'much humus') is an organisation that runs courses throughout the world, but mainly in Latin America, on biofertiliser production and using paper chromatography.

Aerobic Digestion

Aerobic microbes need oxygen, which is pumped into the solution by using an aquarium pump for smaller containers or air blowers (50-80 L/min) for larger tanks. The general term for the products from this process is 'compost tea'. You can suspend a bagful of earthworm castings in a 20 L bucket and aerate for a day or two (generally no more than two days).

On a larger scale, a cereal bag (wheelbarrow-full) of animal manure or shredded weeds and plant material can be aerated by a larger blower or compressor. The drum is open to allow air to escape. Plant material should be finely shredded as much as possible whereas animal manure easily falls apart in water.

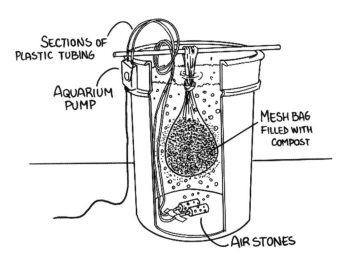

SECTIONS OF PLASTIC TUBING

AQUARIUM PUMP

MESH BAG FILLED WITH COMPOST

AIR STONES

Figure 6.11 Aerobic digestion – makes compost tea.

Compost teas are quick to make, and can be undertaken on-farm in large quantities, but they must be used at once to maximise their effectiveness and there is an energy cost for the aeration. Again, you will need to strain the solution so that it can exit any spraying equipment you use.

Both compost teas and biofertiliser help build soil fertility, and these products can be made from most organic wastes.

Natural Intelligence Farming

A good example of biological farming has been demonstrated by Western Australian farmers Ian and Dianne Haggerty who have called their system natural intelligence farming.

They have demonstrated success by combining natural processes with the latest in technology to produce the highest quality grains, meat and wool, while regenerating the landscape's natural biological functions and water cycles. The Haggerty's chemical-free grain is in high demand.

Instead of chemical fertilisers, pesticides and herbicides, the couple have adapted their large machines to spray a combination of worm juice and compost extract to coat the seeds, enrich the soil and boost the plant's immune system. The key to natural intelligence farming is not to hinder or obstruct the interactions that support and inform the relationships that occur between the soil, plant seeds and roots, microorganisms, and the ruminants that feed on the plants.

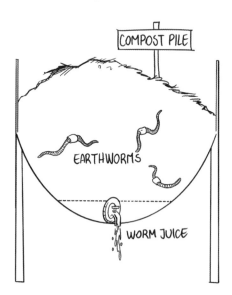

Figure 6.12
Worm juice is a nutrient-rich product.

Final Remarks

There are ample opportunities for anyone wishing to adopt some or many of these practices. If we can develop systems that do not rely on any additional supplementary feed, improve our stocking capacity and general productivity, contribute to mitigating climate change, restore degraded landscapes and provide a fair income for the farmer, then let's all move in that direction.

7

Animals

This chapter is not about breeds of cattle or which varieties of pigs and chickens we should keep, but rather an examination of the roles and functions of animals in ecosystems and on the farm, and how they can be integrated in the farming operation. Cropping alone (even done locally and organically) relies heavily on inputs, often from animals but if not animals then external inputs. Mixed cropping and livestock is common practice throughout the world, yet there is so much more potential for these types of systems.

The Animal Dilemma

One barrier to mainstream uptake of regenerative agriculture is the issue of animals. I should make my position clear before we start: Animals are important on farms, regenerative grazing enables the responsible use of animals as human food sources and the management of grasslands depends on animals, like it has for millennia. However, I also want to present both sides of the story, without too much bias, and in a balanced way. So, let's look at some of the issues about animals before we offer solutions. The four areas of concern relate to environmental damage, the clearing of forested areas for new pastures, greenhouse gas emissions and the treatment of animals.

Many people (not just vegans and vegetarians) take issue with the poor treatment of some farm animals, their slaughter and then consumption by other members of the public. However, if all animals are treated with respect and they have a good and happy life then the role of animals in a farming system can be acceptable by the majority of people. What vegans and farmers probably wholeheartedly agree on is that factory farming of animals has to stop. As we have mentioned many times already, our ability to grow food for vegans and non-vegans may very well depend on rebuilding soil. After all, there is no life without healthy soil, regardless of how animals are treated, so the focus on cropping or livestock agriculture, as a whole system, embraces the living soil as the key to all life.

A growing number of farmers want to become organic and some of these go further to only have a plant-based agricultural system on their farms. You can make compost from just plants and as long as you use techniques that replace the work that animals do on a farm you can be successful. Regenerative agriculture embraces a whole range of strategies and practices that result in reduced fertiliser use, increased biodiversity, planting trees and shrubs, greater soil health and better environmental outcomes, all without using animals.

Good agroecological practices, reforestation and agroforestry can also sequester a lot of carbon and build soil fertility without livestock or fewer livestock, with much lower land use – and without the production of large amounts of methane. With less inputs, herds can be reduced, as smaller numbers is all that is required to meet the bottom line of profitability. On the other hand, reducing livestock numbers can also cause land degradation. Land degradation can occur from both under-grazing and overgrazing, so as I have been saying quite a lot, it's animal management that could be the problem, not the animals themselves.

Population growth and the 'westernisation' of diets means that consumer demand for meat and dairy is likely to increase. Global beef and dairy consumption is still quite high and it generates more greenhouse gas emissions than all the world's cars, threatening long-term climate targets. Cattle are responsible for 9% of all human-induced greenhouse gas emissions, or 4.7 gigatonnes a year. The average large ruminant produces 250-500 litres of methane a day and while smaller animals produce less, there are about three and a half billion ruminant animals on the earth at present. Methane has a short life in the atmosphere: it is changed into carbon dioxide after about a decade or so, but its impact on global warming is far greater than carbon dioxide.

The bulk of emissions generated by cattle, sheep and goats is methane (CH_4), although farms and animals also produce carbon dioxide (CO_2) and nitrous oxide (N_2O). Nitrous oxide is also produced from fertilisers and animal wastes that pass into the soil. Nitrous oxide traps much more heat in the atmosphere than carbon dioxide and methane, and is therefore more of a concern amongst scientists and governments.

In many countries, direct livestock emissions account for the majority of greenhouse gas emissions by the agricultural sector and this makes livestock the third largest source of greenhouse gas emissions after the energy and transport sectors.

Increasing meat production would likely increase the clearing of tropical and temperate forests, wetlands and savannas. This deforestation, driven by the high land requirement for cattle and other stock, already releases huge amounts of carbon into the atmosphere. The concern is not merely about current consumption levels but projections of increasing global meat consumption in coming years. The message that we can eat as much meat as we may want is fraught with danger.

Advocates for reduced meat consumption acknowledge that food ethics are complicated and often personal. Livestock, and especially large animals such as cows, require much more land than most crops consumed directly by people. For example, beef requires about twice as much land per gram of protein as chicken and pork, and twenty times as much land as the equivalent amount of protein from beans. Crops are about twice as efficient at producing protein if consumed directly rather than as meat from poultry or pigs.

Getting consumers to switch 1 kg (2.2lbs) of protein consumption from beef to pork or chicken could reduce the associated CO_2 output by at least 80%. Interestingly, there is a general global trend of reducing beef consumption (mainly because of cost) and greater consumption of chicken and pork. The issue is that regenerative agriculture beef is still only a small fraction of all farm cattle which produce greenhouse gases. Some data on the carbon footprint of various sources of food is shown in Table 7.1, and farmers could consider other types of stock and farming enterprises besides beef. Beef can be notoriously non-profitable as markets fluctuate.

Table 7.1
Food carbon footprint

Animal or plant	kg CO_2/kg matter
Beef	16-19
Lamb	20
Pork	5
Chicken	2.5
Cheese	10
Milk	1.1
Potatoes	3
Lentils	1
Most fruit and vegetables	<2

Farmers, agricultural producers, food companies, restaurant chains, consumers and policymakers all have a critical role to play to curb stock-related emissions. Recently, certain feed additives have been shown to significantly curb cows' methane emissions and free up more energy for muscle and milk production. Along with selective breeding, carbon offsets and various sequestration programmes farmers have a suite of strategies to work towards sustainability.

Damage to the environment occurs when stock compact the soil, remove too much vegetation, and foul waterways. When the existing rainfall decreases or becomes ineffective, it may not be absorbed in the landscape or it runs away, then the problems are compounded. Desertification may even start to occur, especially if poor grazing management is practiced. Overgrazing can contribute to desertification as plants are grazed beyond recovery. It is not that important how many animals are grazing in a particular area (too many = overstocking) as if those are moved before too much vegetation is removed, no long-term harm occurs to the vegetation cover.

While switching from grazing to more intensive feeding systems might lead to better yields and lower emissions intensity per unit of cattle, it is not that simple. While people recognise that there is a difference between confined animal systems and grass-based livestock systems, many do not consider the crop and livestock opportunity that regenerative ggriculture offers.

With an appropriate grazing regime, high levels of soil carbon sequestration more than cancels out greenhouse gas production. Soil carbon sequestration may decrease after several decades of soil neglect and eventually approach zero, but more and more work is being undertaken to ensure sequestration can continue for centuries. It has been going on for centuries in many parts of the world where animals and grasses and other plants live together. An effective, well-managed regenerative grazing operation has the potential of sequestering more carbon than that being emitted.

Over time, if CAFO (Concentrated Animal Feeding Operations or 'factory farms' as they're more commonly known) beef production can be reduced, as it should, there should also be considerable room for increased regenerative agriculture pasture production. This is provided that we collectively achieve lower levels of per capita meat consumption. Individual dietary choices, when combined with more sustainable agriculture implementation, can have a profound effect on the global climate outlook.

Uses and Functions of Animals

Farm animals are typically herbivores – they eat plants (fodder, forage) to obtain the energy they need for survival. Farming indigenous herbivore animals like buffalo in North America, alpacas and llamas in South America, deer in Europe and kangaroos in Australia has not been successful so farmers rely on introduced cattle, goats, sheep and other stock. These have specialised stomachs and digestive systems (they are called ruminants) to break down cellulose and other tough plant components.

Some of the common products and functions of animals are shown in Table 7.2. We can't do without animals altogether, and this list is certainly not complete as animals have many more uses and functions.

Table 7.2 Uses of animals

Products from animals		Function of animals
honey	meat	transport
wax	eggs	ploughing and work
leather	feathers	odour detection
silk	bone meal and fertiliser	companions (pets)
manure	pearls	weed and vegetation control
wool and hair	bone and horns	vermin control
milk	gelatin	protection
mohair, cashmere	hides and skins	cycling of nutrients

Ruminant animals that mow grass and other plants close to the ground (grasslands, pasture) are called 'grazers', while those that chew on leaves, stems, fruits and woody twigs of the larger shrubs and trees are termed 'browsers'.

Even though grazers and browsers have their own niche in the ecosystem, both groups of animals can interchange their roles to some degree. Many grazing animals can be seasonal browsers. Sheep may nibble on pasture grass most of the time, but if the grasses start to dry out and shrubs and trees are present, they will eat the leaves and branches as well.

This is typical of most animals – if their preferred food is not available they will eat something else. Most grazers will also browse and even ringbark trees if their diet lacks certain minerals.

Common farm animals that graze on low-lying plants include sheep, geese, rabbits, horses, donkeys and cattle. Farm animals that are browsers include goats and deer. Non-farm animals such as giraffes and camels also fit into this category.

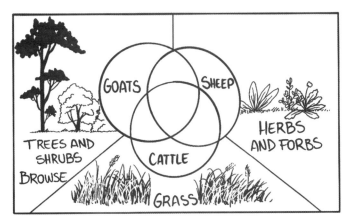

Figure 7.1 Different animals exploit different food sources.

Figure 7.1 shows how there can be much overlap between grazers and browsers. There is a lot of land that is not suitable for growing fruit and vegetables but is suitable for growing grasses and other plants. If we have these vast areas of grasslands we need animals to graze and manage them.

Browsing animals can have an advantage over grazers when grasses die off, as they can turn their attention to woody plants. Grazers have an advantage as there is often more nutrition in the young, rapidly-growing grasses than in old mature shrubs and trees, which tend to be very woody. Grazing animals do not seem to stress plants and stop growth. Plants do shut down for a while but they will recover to allow required crops to get up and become established.

As we have already discussed, the secret to farm success is good grazing management, and this may include high density, short duration, long plant recovery and other planned grazing strategies. The grazing referred to as cell grazing, planned grazing, rotational grazing and time control grazing basically all mean the same thing. It is all about moving stock at the right time, but if we have set stocking (animals roam and stay in the same paddock) then there is unnecessary grazing pressure and problems arise.

Poor grazing management leads to soil decline. Some farmers do not allow stock to remove more than 60% of the above-ground leaf material as it has a severe effect on recovery and also on the soil biomass (organisms). Eating, not walking, does the damage to grasslands. Poor grazing practices and no recovery time kills plants. Short grazing, with long recovery is best. If 50% of the plant leaves are grazed there is little impact on root structure. If say 70% is grazed then roots are seriously affected, stop growing or even are shed by the plant. Some farmers see animals as the keepers of grassland and essential tools and strategies to the management of cropland. Grazing habits encourage new growth and this is an integral part of any ecosystem.

Normally you may need at least six months' recovery time, enough time over a couple of seasons to get the grasses and plants to regrow to an acceptable size. However, this does depend on the farm, climate and land. Some farmers might rest the paddock for 12 to 15 months and others somewhere in between. In fragile and low rainfall systems plant recovery may take a few years after grazing. There is no one size fits all scenario. Some farms can respond and recover after a few months, others far less frequently.

Moving stock every day or even several times a day can be very time-consuming unless you have electric fencing with solar-powered gate releases on a timer that enables stock to move by themselves from pen to pen. Sheep need to be run at high densities so that they move as a mob through the paddock properly. If there are only a few sheep they tend to stay in some areas and some of the paddock is not touched. But, it is no use having intense stock density if you don't have the groundcover.

Another consideration is using multispecies of animals grazing together. It is common to see cattle and sheep or sheep and goats in the same paddock, but any combination of animals might work and complement each other. Many regenerative agriculture farms focus on beef production, but there is a great need for regenerative grass and pasture systems for lamb, pork, poultry, deer and buffalo, for example.

Figure 7.2 Multispecies grazing.

Many livestock systems damage the environment. The way we sometimes hold livestock on farms is not natural and the fodder and food producing systems are not natural ecosystems either. Animals are well-known as compactors of the soil surface. Compaction occurs when the soil structure changes, and soil colloids lose their electrical charges and then disperse. A horse or cow might weigh up to 1 tonne so it's like driving a car over the paddock all the time, squashing and damaging plant leaves, compacting soil and closing air pore spaces. The result of compaction varies from shallow hardpan (from animals) or deep hardpan (from heavier machinery). Thankfully, in a regenerative farm system the soil microbes near the surface can ameliorate the compaction problem when the animals are moved to another area. Generally, soil compaction caused by animals is not long-lasting. Furthermore, there may be little compaction issues if we rehydrate the landscape, have lots of plants and increase soil cover.

Conventional grazing does work: this is where animals are free to roam anywhere over a fenced paddock, eating what they like (the most nutritious weeds and grasses) with little human intervention. Stock tend to fatten up well, and when the paddock is exhausted, stock are moved. But the problem with this practice is that the stock are selective, leaving unpalatable plants alone (such as weeds) which increases these types of plants at the expense of better quality feed.

Needs of Animals

At the onset, the message here is simple: let animals exhibit their full range of natural behaviours. Animals get stressed easily (anxiety) and they tend to not do things that are unfamiliar to them. It's about the pigness of pigs and the cowness of cattle – respect animals and what they want to do and exhibit.

All animals have the basic needs of food and nutrition (protein, fat, fibre, minerals, electrolytes and vitamins), water, air (oxygen), shelter and mates, and if we can interfere and manage animals, then husbandry can ensure optimal health and well-being. To maintain optimum health, all animals require a balanced diet, receiving adequate levels of all nutrient classes on a daily basis.

When it comes to animals, not all breeds are the same. Some breeds have different nutrient requirements and tolerance: highly selected animals are sometimes not able to express their productive potential and are much more susceptible to develop different metabolic disorders.

Keeping stock locked in a small area has its problems – the most important being water. Stock drink lots of water and if their water supply

Figure 7.3
The basic needs of farm animals.

is not moveable along with them, they have to return to where the water trough or dam is, several times a day. Besides moveable electric fences to corral animals you may need a moveable water supply or a number of watering points in each paddock. This might involve kilometres of pipework, hopefully buried so stock don't damage it, with taps or valves at various points where a trough can be connected. Ideally, a number of troughs all set up strategically along the length of a paddock is best, but may be cost-prohibitive.

Food and nutrition is paramount to animals. Grassland is recognised as one of the best ways to produce meat and dairy, but it is not sustainable if we have to feed all stock with plants that require lots of fertilisers or we import feed from elsewhere. It is also ridiculous to grow a crop, harvest the seed and then feed it to sheep or cattle.

Figure 7.4 Water sources can be fixed or portable.

By now it should be clear that farmers need to grow forage and autumn feed (even an oat crop) and fodder (acorns) so that animals have variety and choice.

Food for stock also includes minerals. Some of the functions of minerals and trace elements were covered in Chapter 4 (p.65) on plant function and it is well known that trace minerals (cobalt, copper, iron, iodine, manganese, molybdenum, selenium and zinc, among others) are required for the normal functioning of basically all biochemical processes in organisms.

Minerals and trace elements are crucial for animal health as they are essential constituents of skeletal structures such as bones and teeth, they play a key role in the maintenance of osmotic pressure, and thus regulate the exchange of water and solutes within the animal body, serve as structural constituents of soft tissues, are essential for the transmission of nerve impulses and muscle contraction, play a vital role in the acid-base equilibrium of the body, and thus regulate the pH of the blood and other body fluids, and serve as essential components of many enzymes, vitamins, hormones, and respiratory pigments, or as cofactors in metabolism, catalysts and enzyme activators. Some functions of selected trace elements and the effects that happen if they are deficient are shown in Table 7.3, and information about other minerals is easily found.

Table 7.3 Functions of some trace elements

Element	Functions	Deficiency symptoms
Copper (Cu)	Body, bone and wool growth	Rough coat in cattle or steely wool in sheep, poor growth and body condition, diarrhoea
Selenium (Se)	Growth and prevention of white muscle disease in lambs and calves	Poor growth, reduced wool production, stiff legged gait
Cobalt (Co)	Production of vitamin B12 in the rumen	Poor appetite, weeping 'rheumy' eyes, scaly ears, anaemia and decreased milk production
Iron (Fe)	Red blood cells, resistance to disease, formation of DNA	Anaemia, reduced growth, loss of appetite, and increased infection rates
Zinc (Zn)	Enzyme production, essential for growth and reproduction	Sub-normal growth and fertility, crusty proliferations, cracking of the skin and loss of hair

Requirements for minerals are hard to establish and most estimates are based on the minimum level required to overcome a deficiency symptom and not necessarily to promote productivity. Supplements used in intensive animal feeding operations generally have concentrations several times greater than minimum requirements, and manufacturers often raise mineral concentrations in feeds far above the actual needs of livestock.

High levels of mineral supplies do not normally cause animal health issues as there are relatively high security margins for most trace minerals and in most domestic species. However, in some cases, supplements have a high heavy metal content. Mineral supplements generally contain trace residues of toxic metals, especially of cadmium (Cd) and lead (Pb). These heavy metals have bioaccumulative properties and can accumulate at very high concentrations in the liver. There is much evidence that hepatic and renal cadmium accumulation was significantly higher in pigs from intensive systems, so this is another reason why the use of intensive feed lots must stop.

Furthermore, this growth of the intensive livestock sector stresses many ecosystems and contributes to global environmental problems. This is because a significant proportion of minerals that are given to livestock are excreted in urine and faeces, and then excessive mineral supplies and their loss in elimination from the animal pollute the environment.

Another issue is that the physiological role of many trace elements is often underestimated or not known. Many trace minerals often have multiple roles and their antagonism or synergy is little understood. Even so, a slight deficiency of trace minerals can cause a considerable reduction in performance and production.

In sustainable systems the main source of trace minerals for livestock is the soil itself and the waste materials from the farm which are recycled on to the pasture. The available mineral concentrations in the different soil types are dependent on the concentrations in the parent rock and on the chemistry of the soil, and as a consequence, forages in some circumstances may be deficient or imbalanced in some trace minerals. In rare cases, acute deficiency or toxicity may occur and remedial actions may be complex and difficult.

Trace mineral concentrations are generally higher in the surface soil and decrease with soil depth. In spite of the high concentration of most trace minerals in soils, only a small fraction is available to plants, and soil pH is one of the most important factors affecting it. Trace mineral concentrations in pasture also varies seasonally. Leguminous species are generally much richer in macro-elements than grasses growing in comparable conditions.

However, a great number of soils around the world show mineral deficiencies or imbalances to a greater or lesser extent and trace mineral distributions are considered one of the main farm problems. Feed the soil to feed the plant is absolutely essential, so that all nutrients removed in farm-produced products are replaced.

We must guarantee animal health and optimise animal production together with a respect for the environment and to get animal products with an adequate trace mineral content, so a good diagnostic plan is essential to correctly quantify them.

At present the main strategy for improving trace element status in livestock is to analyse the soil, forages and possibly other feeds, and then decide on the best course of action.

Typically, this action is mainly through mineral blocks and other similar strategies. However, mineral blocks and use of trace element injections are just a band aid approach. It is an immediate fix but not really addressing the problem and finding a long-term solution. In cases of animal deficiency or toxicity, it is more economical to treat animal health than to amend the land by the application of fertilisers and trace elements (soil additives), but this strategy is short-sighted. If the soil is made right then everything else falls into place.

For this reason, there is a growing interest in proteinated or chelated trace minerals. In this form, the trace minerals are chemically bound to a chelating agent or ligand, usually a mixture of amino acids or small peptides. Many minerals, such as copper, manganese and zinc, are carried on organic complexes within the soil. This makes them more bioavailable and provides the animal with a metabolic advantage that often results in an improved performance.

In recent years, however, perhaps in response to concerns about intensive farming methods we now see movement towards organic and other sustainable production systems, but where mineral supplements may still need to be added. Here, a farmer needs to know if there are mineral deficiencies in the soil and plants so this can be addressed for optimum animal nutrition. Minerals must be provided to livestock in optimal concentrations and according to requirements. Trace elements are required for animal growth and production, so in plants low trace element levels can cause nutrient deficiencies in grazing animals. The application of farmyard manure and inclusion of plant species containing high concentrations of trace minerals in the pasture can increase the supply of trace minerals to grazing livestock.

Within domestic animals, there are marked variations in their tolerance to the increased levels of dietary minerals. For example, higher levels of copper can be tolerated by pigs and cattle, but not sheep where it is toxic and leads to chronic hepatic accumulation. Too much causes chronic copper poisoning and this affects liver function and ruptures blood cells.

Associated with copper are the interactions of high molybdenum and iron concentrations, which can induce copper deficiency in livestock, and high levels of calcium in the feed inhibit the uptake of zinc. Copper deficiency may occur in animals due to pasture type, as grasses have a lower copper level than clovers and legumes, while cobalt, and therefore vitamin B12 deficiency can be seen in association with heavy liming and applications of superphosphate. It is not hard to understand that herbage mineral concentrations are often inadequate to meet the dietary requirements of stock. Getting the right balance of all minerals is challenging and most efforts will be trial and error, with a bit of science thrown in.

The availability of many minerals is dependent on the acidity or alkalinity of the soil. At soil pH values below 6.5 the availability of molybdenum (Mo) and selenium (Se) is reduced and the availability of iron (Fe), manganese (Mn), cobalt

(Co), zinc (Zn), and boron (B) is increased; the opposite is true at soil pH values above 6.5. A high soil pH can also cause boron deficiency, resulting in the poor establishment of clover or other nitrogen-fixers which is essential for the provision of nitrogen in organic systems. Severe zinc deficiency results in poor pasture growth, which is due in part to the poor utilisation of nitrogen within the plant when zinc is low. Figure 7.5 shows the changing availability of some minerals due to soil pH.

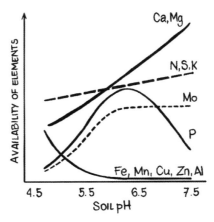

Figure 7.5 The effect of pH on nutrient availability.

Besides soil pH, other factors affect minerals. For example, selenium, iodine and boron are relatively easily leached by rainfall, and there are synergies and antagonism between elements which further affect availability. Some of these ideas were discussed in Chapter 4 (p.65) on plant function.

Most trace minerals are present at higher concentrations in cereals than in grass forages, and while nitrogen-fixing legumes may have higher levels of some trace elements, these may not be available to animals. This is due to substances, such as phytate, which can bind with elements and thus reduce the biovailability to the animal. Table 7.4 shows some levels of minerals in various plants and it is clear that these levels vary markedly, but scientific research into individual herb species, their mineral concentrations and their availability is still limited.

Soil ingestion is an under-rated mechanism for animals to obtain nutrients. For example, very little cobalt is taken up by a plant; therefore, it is mainly by soil ingestion that sheep and cattle can fulfill their needs of this trace mineral. Ingested soil can occasionally supply greater than 60% of trace minerals to the animal, including copper, iron and zinc. In pigs, the rooting behaviour from their day of birth implies soil ingestion and a regular iron supply from the soil.

On the other hand, soil ingestion may be an important pathway for animal exposure to toxic trace minerals. For example, the relatively low accumulation of lead in pasture plants when compared to a high degree of soil contamination suggests a potential soil-animal toxic metal exposure in polluted areas. This is true of most other heavy metals such as cadmium and arsenic.

Table 7.4 Nutritional value of some fodder plants

	Calcium mg/100g	Magnesium mg/100g	Phosphorus mg/100g	Iron mg/100g	Sodium mg/100g	Zinc mg/100g
Pigeon pea	80	6	108	0.3	330	3
Amaranth	265	1320	480	46	17	15
Drumstick tree leaves	76	42	29	690	650	310
Field bean	42	135	355	7	0.2	2
Wheat grain	40	170	470	4	30	4.5
Oat grain	100	160	340	8	90	2

Soil ingestion will increase with low amounts of forage available, various seasonal conditions, high stocking rates, root intake and loose soils. Soil ingestion when grazing can significantly contribute to parasite exposure. While we have discussed the stocking density of animals in paddocks, high stocking rates are linked with increased parasite loads; lowering the animal density serves two purposes: it reduces the amount of manure in a given area and the residual grazing height of the forage is often much higher, which significantly reduces the probability of parasite infection (80% of parasites live in the first five centimetres of forage above-ground). Particular forage plants have a role to play too. Chicory, a plant with low tannin content but high soluble carbohydrate and minerals, may enable stock to tolerate parasites, and it has been shown to offer animals some protection from internal parasitism. Again, farmers must move stock before plants are too short and are slow or unable to recover. It's all about adaptive planning – destocking before there is critical damage to the soil.

A number of seed houses are now offering seed mixes which include a range of grasses, clovers, trefoils, and herbs. While these mixes are generally significantly more expensive and take longer to establish, the indications are that they are much longer lasting. If they also result in better trace element status in the livestock it is likely that the extra cost will be more than offset by a combination of less frequent re-seeding and better animal health. Again, the components of pasture that we need to be concerned about are plant density, number of tillers (shoots) per plant, the height of the grass, and species composition.

Integrated Animal Systems

It is ironic that, historically, plant and animal agriculture developed simultaneously in many places around the globe, but the current separation of animal production from cropping systems is relatively new.

In an integrated system, livestock and crops are produced within a managed operation. The waste products of one component serve as a resource for the other.

For example, manure is used to enhance crop production; crop residues and by-products feed the animals, thus contributing to improved animal nutrition and productivity. The farming system is essentially cyclic (organic resources, livestock, land, crops), so management decisions related to one component may affect the others. It's also the permaculture principle that the residues, wastes and by-products of each component serve as resources for other components. Nothing is wasted.

An integrated farming system consists of a range of resource-saving practices that aim to achieve acceptable profits and high and sustained production levels, while preserving the environment through prudent and efficient resource use and minimising the negative effects of intensive farming. The increasing pressure on land and the growing demand for livestock products makes it more and more important to ensure the effective use of feed resources, including crop residues.

Using the correct management of crop residues, together with an optimal allocation of scarce resources, leads to sustainable production. Growing fodder legumes and lots of other plants and using them as a supplement to crop residue is the most practical and cost-effective method for improving the nutritional value for animals, particularly during dry periods.

Integrated, mixed farms are systems that consist of different parts, which together should act as a whole. We recognise that the most important outcome is to achieve increased production in mixed systems, together with the awareness that crops and animals have multiple functions, and that their integration serves to make maximum use of the resources.

In these crop-livestock systems, cropping provides animals with fodder from grass and nitrogen-fixing legumes, weeds and crop residues. Animals graze under trees or on stubble, and they provide nutrients in manure and urine for crops.

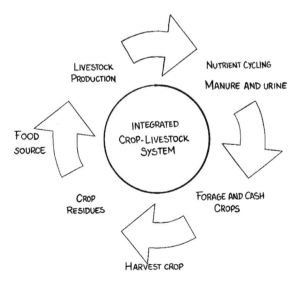

Figure 7.6 The cyclical nature of an integrated crop-livestock farming system.

One of the main permaculture principles is to mimic nature in all its biodiversity. This is why we utilise as many species of useful plants as possible in livestock feeding systems design, each with different growth habits and functions. Such diversity boosts productivity and adds resilience to our livestock feeding systems.

The overall benefits of crop-livestock integration can be summarised as follows:

- Economic, through product diversification, and higher yields and quality at less production cost. It is efficient and economically viable because grain and other cash crops can be produced in four to six months, and pasture formation or continuation after cropping is relatively rapid and inexpensive.

- Ecological, through the reduction of crop pests (less pesticide use) and better soil erosion control. Combining ecological sustainability and economic viability, the integrated livestock-farming system maintains and improves agricultural productivity while also reducing negative environmental impacts.

- It helps improve and conserve the productive capacities of soils, with physical, chemical and biological soil rehabilitation. Animals play an important role in harvesting and relocating nutrients, significantly improving soil fertility and crop yields. However, animal manure alone may not meet crop requirements, even if it does contain the kind of nutrients that plants need. Alternative sources of the nutrients may need to be found. The temptation is for farmers to use chemical fertiliser instead of manure because it acts faster and is easier to use, but as we have discussed already, this often leads to other problems.

- It results in greater soil water storage capacity, mainly because of biological aeration (earthworms and other soil animals) and the increase in the level of organic matter through stubble, digested plant remains, manure, urine and decaying trampled forage.

- It provides diversified income sources, guaranteeing a buffer against trade, price and climate fluctuations. This alone is enough motivation to adopt integrated, multi-species farm operations.

Sustainable intensification of land-use practices has never been more important to ensure food security for a growing world population. With good management, cover cropping and crop-livestock integration under no-till systems, unexpected synergies between plant-animal diversification and complex agroecosystem functions are possible. However, farmers need to have sufficient access to knowledge, assets and inputs to manage this system in a way that is economically and environmentally sustainable over the long term. This includes better livestock management that is needed to safeguard water, which is the pillar for all farming.

Care of Animals

It's a myth that farmers don't care about their animals, soil and plants on their farm or the environment. They do, it's just that some don't know how they can do it better. I once read that "farmers are good people trapped in a bad system", so we would hope that farming animals differently would not only benefit the health and well-being of the animals but also benefit the farm.

Globally, livestock production has shifted towards more intensive, large-scale and commercially oriented, specialised systems. Animals are typically fed anything in a concentrate feed that might fatten them up, even though there has been some research suggesting that if you remove all supplementary trace minerals from a fattening ration for some animals it might only have minor effects on performance and carcass quality, and not all of those effects are negative.

We all know that feed lots and confined animal feeding systems contribute to a range of health and environmental problems, and these systems rely on low nutrient density feed, antibiotics and other supplements, and result in water pollution and soil deterioration. There's the rampant overuse of antibiotics in raising livestock, especially in crowded feedlots (CAFOs).

Figure 7.7 Animals need considerable care.

Besides antibiotics, stock are subject to the chemical control of pests, parasites and afflictions, dosing of animals (drenching) for de-worming and dunking (dip) them in a bath of insecticide and fungicide to control external parasites. These chemicals would ultimately affect soil life too, and you could eliminate drenching and other health conditions if you just made free choice humates and licks available for animals. Animals can determine what minerals they need and what levels are required, and I contend that in a well-managed regenerative farming system animals do not suffer from excessive parasite problems.

Keeping and using animals on farms can be both rewarding and profitable, but it is essential that enough time and resources are devoted to their health and well-being. In particular, these kinds of things should be undertaken as the minimum care:

- Provide abundant, clean water daily or as required.

- Provide sufficient variety of food (each animal has different diet requirements) to meet their nutritional needs.

- Keep the living area clean for those animals that may be penned or occasionally held in enclosures.

- Provide the opportunity for the animal to be themselves, such as walking and running in open spaces.

- Most animals need mates, so be sure to have at least two of a breed or variety, or just other sorts of animals they can interact with.

- Interact with your animals regularly, not only so they get used to your being close-by, but also so bonds can form.

- Keep young and susceptible farm animals in an appropriately-sized enclosed area at night to decrease their exposure to predators.

- Use elevated feeders and water troughs, and store feed off the ground in sealed containers. This will help keep pests like mice and rats at bay.

- For animals in pens for any reason, properly dispose of soiled bedding and spoiled or uneaten food so it doesn't make your animals sick.

- Good hygiene is crucial. Clean farm animal enclosures frequently, and wear personal protective equipment (PPE) while cleaning, assisting an animal in giving birth or doing any activities that would involve touching bodily fluids from animals. PPE might include overalls, boots, gloves and a mask.

- Monitor your animals' health. Talk to your veterinarian if you suspect any animals are sick. Schedule routine veterinary examinations to keep your farm animals healthy and to prevent infectious diseases and internal and external parasites.

Besides the list above, changes to farm infrastructure contribute enormously to animal well-being. For example, in adverse weather, especially frost, snow and extreme cold, stock can die. This is why you need shelterbelts where stock can shelter against the wind and huddle together to conserve heat loss. Stock do graze through snow and they do prefer to graze rather than being locked up in a pen or in a shed. The ideas behind agroforestry, shelterbelts and microclimates are covered in previous chapters.

Final Remarks

The broader community desperately needs farmers that can make a living by practicing agriculture that heals the earth and communities. And farmers need the rest of the climate change community to support them if they are going to be able to push back effectively against environmentally and socially destructive corporate industrial farming that is killing the countryside and the planet.

Some regenerative agriculture advocates tell us that, if produced correctly, we should be able to eat as much beef as we like. Unfortunately, this unlimited meat message, mainly promoted by the industrial meat industry, is misguided. Currently, people in industrial and affluent countries consume high amounts of meat, milk, and eggs. We must eat less, and eat foods produced by environmentally and socially sound means such as agroecological, organic or regenerative agriculture methods.

Ultimately, we need to change from being stock managers to land managers. It is not animal numbers that cause overgrazing and desertification. It is the management of these animals on that land that is the issue, and this includes the soil and all that it contains. Mineral deficiencies or imbalances in soils and forages account partly for low animal production and reproductive problems, and highly efficient grazing management is required to get an adequate mineral status.

The way to deal with trace mineral needs in organic farming will mainly depend on the own-farm husbandry and management practices. Livestock practices are highly standardised in intensive farming, but are very variable in organic systems. Up until now there has been little trace mineral supplementation allowed in organic systems and there may need to be. It's all about balance. Organic farming with crop rotation ensures good soil fertility and plant health, but consumers generally assume products from organic farms are more healthy and nutritious, and this is not always true if the farm is located in a trace mineral deficient area.

Animals can generate extra profit, and with locally owned processing and marketing, it can be a very good way to add income for farmers. Often, it is hard for farmers to make a good living on commodity crops alone, as the overproduction and resulting low prices built into the current market system frequently leads to unsustainably low prices. This is why we also need comprehensive policies that support fair crop and livestock prices, break up monopolies, control supply, and foster ecological practices.

8

Integrated Pest Management

Many farmers and growers already know a little about, or even embrace, integrated pest management (IPM). The principles of IPM are straightforward, using a five-pronged approach focussed on prevention, cultural, physical, biological and lastly chemical strategies to avoid, suppress, minimise and control pests and disease.

Regenerative agriculture and IPM go hand-in-hand. Many strategies are required to manage all types of diseases and pests and some of these are discussed in this chapter. Regenerative agriculture also uses various approaches to improve soil health, ecosystem function and environmental harmony, and is really the only way we should farm. What needs to be understood is that regenerative agriculture is more than improving soil health and offering a direction toward true sustainability. In its purest form, and in our purest understanding, regenerative agriculture does not use chemicals. And this is where the problem lies.

Conventional farmers who want to do things differently, adopting practices such as pasture cropping, holistic management or natural sequence farming, are still able to use chemicals to combat weeds and pests albeit endeavouring to use less of these. One of the challenges that prevent many farmers turning to organic is how to control weeds and pests without resorting to chemicals that tend to be quick acting, relatively cheap and easy to apply.

Unfortunately, in regenerative agriculture circles, the issue about pesticide and herbicide use is mainly ignored and overlooked, with presenters at regenerative agriculture seminars, workshops and courses simply allowing farmers to make their own decisions about chemical use. This is because farmers may be comfortable with their lifetime use of chemicals as tools for greater food production, regardless of the fact that the monoculture farming model that supports chemical use has stopped working and is contributing to sending farmers broke. It also doesn't help if farmers are battered by fluctuating commodity prices and getting low returns for their produce.

Organic farming practices that rely on tillage to manage weeds often aren't effective enough, and soil disturbance causes a whole host of soil problems. Better, holistic strategies are required and must be seen as cost-effective if farmers are to transition.

Furthermore, it is really a conflict of interest if chemical companies jump on the regenerative agriculture bandwagon. Any amount of evidence is available to clearly demonstrate that farm chemicals seriously damage the environment and kill soil life, which drives carbon sequestration.

Regenerative agriculture is gaining momentum, certainly amongst conventional farmers who have historical chemical-dependent operations. Most farmers who are transitioning still rely on chemicals to control weeds, often planting genetically modified (GM) crops that enable chemicals to be used. However, it's a sobering thought that with all the chemicals we have invented and applied over all the years not one pest has ever been eradicated.

Chemical use is a vicious circle. Using herbicides will reduce weeds but they also decrease biodiversity and the number of beneficial insects, so pests increase and you need pesticides to reduce these. Input costs increase as a result. At the end of the day farmers just need to minimise pest damage to below any economic impact. We need successful alternatives.

Weed management seems to be the number one problem for farmers. The occurrence of weeds is taken as a sign of failure, they are the enemy and shouldn't be permitted. However, we have a newfound understanding about the role of weeds in ecosystems as pioneers, as protectors of the soil, of contributing to soil fertility and as plants that attract pollinators or predators, or both.

Weeds happen. Weeds are survivors. You can find them in waterlogged soil, acidic soil, compacted soil, 'gutless' sands (leached soils) and nearly every other soil type category. Weeds are pioneer plants, they are the first to grow. Their job is to cover and protect the soil – that's their niche in ecological terms. If we deliberately plant cover crops to fill that niche or change the soil through amendments and practices, the weeds are out of a job.

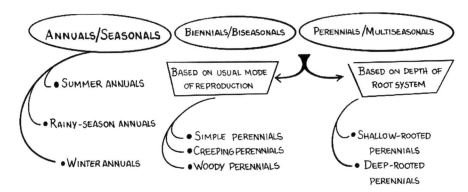

Figure 8.1 Classification of weeds: as annuals, biennials or perennials.

It is important to know your enemy and work that to your advantage. All weed species have their weaknesses and their strengths, usually occurring at distinct stages of their lifecycles or resulting from specific growth patterns. Different weeds present problems at different times of the year, or with different crops. Grassy weeds often require different control measures than do broad-leaved weeds. Correctly identifying the species of weeds that are causing major problems in your fields is critical, to not only choosing but also timing effective control measures. It is valuable to have a good weed-identification book and use it regularly during the season until you are confidently recognising your most common and troublesome weeds.

The biggest challenge is perennial weed control, and invasive perennial weeds require a persistent, multi-pronged approach. Perennial weeds tend to be one of four main types:

- Shallow creeping stems or rhizomes, e.g. blackberry, common nettle and creeping knapweed.
- Deep creeping stems or rhizomes, e.g. kikuyu, couch and nutgrass.
- Tap rooted weeds, e.g. dock and dandelion.
- Tubes, bulbs or corms, e.g. oxalis, cape tulip, watsonia and bridal creeper. Tubers and bulbs are the most difficult to control, especially if the plant is not palatable or is poisonous to stock. In addition, their many storage organs can remain dormant in the soil for a few years making their control and elimination a long-term process.

Ideally, we grow the weeds first, reduce them and then sow the crop. If you are able to use irrigation to germinate the weeds do so, but most farming operations simply rely on natural weather and climate patterns, and irrigation is not an option. The use of precise drip or trickle irrigation technology will reduce general weed growth compared to overhead irrigation, making these ideal for vegetable growing or for fruit and nut trees.

Increasing the crop seed sowing rate by 5-20% can also enable the crop to out-compete weeds (mainly by shading them) and enable more crops to be grown. This may also be necessary as many seeds and young plants do not recover from tractor tyres and machinery squashing them when seed is about to germinate or young plants start to develop. One way to control weed growth is to have highly competitive crops, as a vigorously growing crop is less likely to be adversely affected by weed pressure. However, sowing crop seeds into an existing, established weed patch is not advisable as the weeds will have a selective advantage and outcompete the newly sprouted crop seeds.

Killing weeds is most effective if the weeds have only a few true leaves – the weed plants have only just emerged and are quite young. Young weeds haven't built up storage reserves nor developed extensive root systems and strong stems so they are easier to manage.

Weed seed germination is often dependent on some combination of soil microbe levels, oxygen and carbon dioxide levels, temperature and either light or dark. If the germination conditions aren't right then the weed seed store in the soil remains high. This may pose problems over subsequent years until such time that the weed seed storage diminishes enough so that the few weeds that do germinate have little effect on the crop.

It's all about balance: minimising the negative impact that weeds have on production and tolerating some weeds that make no difference. Understanding how the whole farm operates and the sequence of crops and land uses over time are all considered when weed management is instigated.

Unfortunately, pundits are still strongly pushing a failing system. You have to ask why? Conventional farming, with its heavy use of salt fertilisers, herbicides, monoculture and imbalanced cation ratios could be described as providing conditions for weed enhancement. The conventional farm environment encourages herbicide-resistant weeds that thrive and it is really no wonder that most herbicides are only effective for a few years before a newer, stronger (and more expensive) chemical is needed to control weeds sufficiently.

Similarly, pests are indicative of system imbalance. Ten times more pests are found on conventional farms even with all the chemicals applied (fertiliser use increases insect and disease pressure too), compared to regenerative farms. This makes sense as pests only attack weakened plants high in amino acids and incomplete sugars. Soil pests are at least as significant, in terms of economic damage, as above ground pests.

Insecticides reduce all insect numbers, not just pests. For example, if you use insecticides you may kill dung beetles and other similar insects involved in dealing with animal manure. Very few insecticides are selective, and above-ground there has been a global decrease in bee populations as a result of widespread spraying.

Farmers need more tolerance to pests and disease when they start seeing problems on produce. Generally, the numbers of beneficial insects and those that have some ecological impact far exceeds those who we call pests. Unfortunately, pesticides (and herbicides) cause total destruction of all life. When we spray, everything dies, including the many hundreds of beneficial organisms we rely on to grow soil and maintain a healthy ecosystem.

Principles

Any agricultural endeavour has high pest and disease pressure, simply because you are growing a lot of food and plants that pests eat. There is a huge number of potential pests on farms, which includes disease (due to bacteria, viruses and fungi), weeds, insects, nematodes and invasive (introduced, non-endemic) trees and shrubs.

As we discover more about the very complex relationships and interactions between plants and microbes (fungi and bacteria) that contribute to plant defence mechanisms, the control of pests may become easier. For example, whilst we already know about various fungi and nematodes that play a role in insect control and natural biopesticides in healthy soil and in the plants themselves, our knowledge and understanding of these complex interactions is limited.

The overarching principles of integrated pest management provide some guidelines for what to consider when designing an IPM programme. A brief summary of these is:

1. Know Your Weeds

Knowing the lifecycle of various common weeds that you come across is a step in the right direction. There may be some stage of their lifecycle where they are most vulnerable, and therefore easier to control. There are lots of publications that showcase common weeds in different areas, states or countries. Source some of these and use them to identify weeds in your patch. Watch out for new weed species which may develop into severe problems.

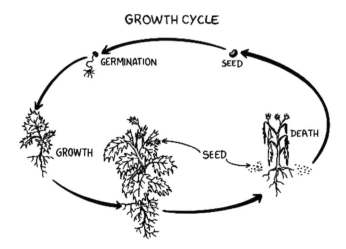

GROWTH CYCLE

Figure 8.2 Typical lifecycle of a weed.

2. Don't Create Opportunities for Weeds to Become Established

This includes using concepts and techniques such as minimum till, no bare ground, tight rotations between one crop and the next and no (unnecessary) soil disturbance. We should aim to interrupt the lifecycle of particular weeds that are adapted to frequent tillage. There are many weeds of cultivation, so modifying control practices helps reduce these. Use a mix of warm and cool season crops so there is always soil cover.

3. Nurture the Soil

As soil changes so do the species of weeds. For example, you often find an increase in native grassland species and a corresponding decrease in weed species if the soil changes to be more loam-like with good structure and texture. When this occurs insect numbers and diversity increases and this is how pests are controlled. Seems strange to believe that you control insect attack by encouraging more insects, but that is the reality. Insecticides wipe out all the beneficials and the predator-prey cycles do not take place. It is similar for fungicides. Fungicides will kill beneficial fungi, such as mycorrhiza, that actually help to control pests by encouraging beneficial microbes and other soil organisms to flourish. Having a robust, healthy soil full of microbes contributes to pest control.

It is not so much about eliminating weeds but rather to develop soils and conditions that produce the lowest weed pressure. This is achieved by encouraging healthy soil conditions with a diverse microbial population. As discussed in the soil chapter (p.27) this means getting the right mix of organic matter, humus, loam, microbes, earthworms and mulch.

We need to aim towards reducing the seed bank of weeds, encouraging seed consumers (conservation biological control) and deflowering weeds so that new seed is not produced. Imported soils and composts may contain weed seed so ensure that these amendments are clean and come from reputable suppliers. If you are making your own compost be sure it is made by a hot composting method so that weed seeds and pests are killed during this process.

While there is some truth in using plants as indicators of particular soil conditions, you should also get the soil tested in a laboratory to give you more concise nutrient and pH levels in your soils. This, in turn, enables you to address any extremes and to choose an amendment to assist in the correction of any pH or nutrient deficiency problems.

4. Guilds

Guilds are discussed in Chapter 5 (p.85) and these are an important component of any regenerative agriculture enterprise. Basically, particular plant-plant and plant-animal associations create synergy and can contribute to pest control. For example, plant types such as *Paraserianthes lophantha* (formerly Albizia) and tagasaste (*Cytisus proliferus*) are useful additions in an orchard as they are nitrogen-fixing, fast-growing, create shelter (nurse trees) and provide windbreak. *Paraserianthes* also attracts ladybirds while tagasaste branches are a snail trap (snails hide in there during the day and can be found and removed).

Companion planting, water sources for animals and allelopathy are just a few other ways to create conditions where weeds are not welcome or seed predators are encouraged. Allelopathy is that phenomenon where plants release chemicals from their roots and some of these inhibit the growth of other plants nearby. Many cereal and other cover crops possess this ability so the overall weed pressure is reduced.

5. Design and Location

Diseases and pests often occur in particular environmental conditions. For example, moist, still air (or soil) encourages fungi on our crops, sheltered and shaded areas provide ideal conditions for bacterial attack and poor, unhealthy soil makes weak plants that are vulnerable to attack. Part of the strategy to counteract these problems includes grouping plants that have the same water requirements, monitoring watering of plants to ensure no excess or shortfalls and the use of suntraps (for tropicals), breezes (for deciduous fruit) and north-facing slopes (for citrus).

6. Balanced Nutrition

The chapter on soil (p.27) was very clear about getting the soil nutrition right and to not feed the plants but rather feed the soil. It has never been about just piling on the animal manure or compost thinking that building up the organic matter in the soil is the key. This often only causes a mineral imbalance of what nutrients are available and which become limited. There needs to be a balance of trace elements, nitrogen-fixing plants, regular plant feeding with foliar sprays and slow release amendments, and the use of good compost and other sources of organic matter.

If relevant to your farm operation, don't over-water or over-fertilise crops as weeds get these nutrients too. Weeds are exceptional competitors, accessing and utilising resources more effectively than some of our crop plants. This competitive advantage means that weeds will always be present, but there are strategies to deal with this, including ensuring no imbalance in nutrients so weeds cannot flourish.

7. Limit Artificial Chemicals

Ideally in a regenerative agriculture operation there should be no insecticides, pesticides, herbicides and fungicides. While chemicals are effective, their use should be severely limited. The secret is to engage many different strategies rather than rely on only one way to reduce weeds and pests.

8. Have a Diversity of Strategies

Use IPM and a range of preventative, cultural, physical (mechanical), biological and, as a last resort, chemical strategies. These are discussed in more detail later in this chapter.

At least try to remember: You don't have too many pests, you just haven't got enough predators to eat them.

9. Increase Biodiversity

If you increase the biodiversity on the farm you end up with many more beneficial insects and other organisms and proportionally fewer pests.

Maintain a long-term balance of diverse crops on a farm, taking into account any necessary soil conservation practices along with crop rotation, livestock requirements, time constraints and market profitability.

In some instances, it might require you to plant disease-resistant varieties if certain pathogens are prevalent in the area.

10. Maintaining and Maintenance

Once your regenerative agriculture IPM programme is in full swing it is time to maintain the system. This might include pruning and shaping, fertilising, irrigating, weeding, as well as sanitation and hygiene, such as using clean seed, cleaning your machinery and making hot compost.

Whatever combination of techniques you use, leave a row or section where no treatment is applied, so you can gauge the effectiveness of your strategies. While few of our endeavours are likely to be so effective that they can replace herbicides and pesticides immediately and completely, they all contribute to a holistic approach to weed management.

Designing an IPM programme

A thorough weed and pest management programme is crucial for success, as many different strategies will be required as one season changes to another, one soil type changes into another or as the types of weeds change through succession.

IPM is not a single pest control method but, rather, a series of pest management evaluations, decisions and controls. In practicing IPM, growers who are aware of the potential for pest infestation follow a four-tiered approach:

Set Action Thresholds

Before taking any pest control action, IPM first sets an action threshold, a point at which pest populations or environmental conditions indicate that pest control action must be taken (for economic, health or aesthetic reasons). Sighting a single pest does not always mean control is needed. The level at which pests will become an economic threat is critical to guide any pest control decisions.

Monitor and Identify Pests

Not all insects, weeds, and other living organisms require control. Many organisms are not harmful (we call them innocuous), and some are even beneficial. IPM programmes work to monitor pests and identify them accurately, so that appropriate control decisions can be made in conjunction with action thresholds. Monitoring and identification removes the possibility that pesticides will be used when they are not really needed or that the wrong kind of control will be used.

Prevention

As a first line of pest control, IPM programmes work to manage the crop or production space to prevent pests from becoming a threat. In an agricultural crop, this may mean using techniques such as rotating between different crops, selecting pest-resistant varieties and planting pest-free rootstock. These control

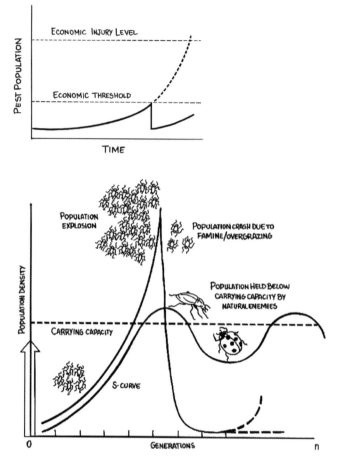

Figure 8.3 (Top) Action threshold. (Bottom) The predator-pest cycle.

methods can be very effective and cost-efficient and present little to no risk to people or the environment.

Control

Once monitoring, identification, and action thresholds indicate that preventive methods are no longer effective or available and pest control is required, IPM programmes can then evaluate the proper control method for both effectiveness and risk. Effective, less risky pest controls are chosen first, including highly targeted chemicals, such as pheromones to disrupt pest mating, or mechanical control, such as trapping or weeding. Should further monitoring, identifications and action thresholds indicate that less risky controls are not working, additional pest control methods would be employed, such as the targeted spraying of pesticides. These would always be the last resort, as chemicals should play a supportive role, not a disruptive one.

To plan an effective weed-control programme, you must integrate a broad spectrum of important factors that include the soil conditions, weather, crop rotations and field histories, together with machinery, markets and specific market quality demands, as well as available time and labour. You must have the ability to adjust your weed and pest control strategies to the unique and ever-changing challenges of each year. Above all, you must be observant, and don't be subject to a knee-jerk reaction to get rid of all pests or weeds at any cost.

Strategies

As we have discussed, preventing the pest from becoming established is the first step and one of the key strategies to control pests is to monitor them. Monitoring can be as simple as walking around the property and observing weeds, insects and signs of disease. You may also use traps to determine which pests are present or you can use the weather forecast to predict possible pest outbreaks.

Traditional IPM takes advantage of all appropriate pest management options and these are outlined below. The emphasis is on control, not eradication. Control methods include the following, written as concise points and not in any order of importance. These apply to a range of farming styles from large broadacre farms to backyard vegetable production gardens, so only some may be applicable to your enterprise.

Cultural Practices

Cultural practices are those things we do – our plantings, our maintenance and our management of our crops.

- Simple hand-picking of insects and snails.

- Select disease-resistant stock and rootstock for the main fruit crops.

- Use plant competition and plant characteristics to control land and aquatic weeds. For example, pine needles inhibit weed seed germination.

- Plant trap crops, such as dill which attracts tomato hornworm and hyssop which attracts white cabbage butterflies. These sacrificial plants are used with the expectation of not harvesting any for yourself. For example, cherries are the favoured food of parrots so plant them some distance from other species in the orchard.

- Use mixed planting in orchards and gardens to encourage predatory species. For example, buckwheat, parsley and yarrow attracts hoverflies; zucchini and other marrows attract ladybirds; fennel, coriander and tansy for lace-wings. These beneficial insects predate on aphids, leafhoppers, mealy bugs and many other insect pests. Plant insectaries to attract beneficial insects.

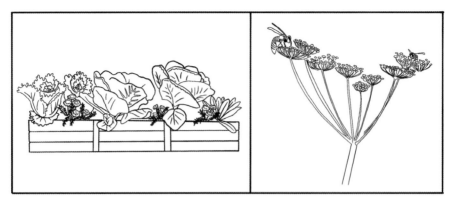

Figure 8.4 (L) Companion planting disguises crops.
(R) Fennel is an insectary plant – it attracts predators.

- Grow herbs and other plants in companion planting. While some companion plants attract predators, others may influence the soil and biota, and thus affect crop growth, nutrient availability and disease resistance.

- Practice sound management and husbandry to discourage soil and leaf pests. For example, remove fallen or diseased fruit and crop residue, cover composting material and use crop rotation.

- Use quarantine and exclusion zones (buffer strips) as the first line of defence. Buffer strips are discussed on pages 106-109. Quarantine imported soils and plants for as long as possible.

- Grow vigorous forage plants to out-compete and smother weeds. Fast-growing, healthy plants might also be able to withstand insect and other pest attacks and minimise the damage to themselves.

- Select plants or animals for specific traits. For example, Brahman-infused breeds are stocked in tick-prone areas as they show high levels of resistance to tick attack, and plant low chill fruit trees to miss the fruit fly season.

- Water and fertilise (irrigation and nutrition) to maintain plant health. Hand watering may be appropriate in small-scale operations.

- Pruning and/or mulch to maintain plant health and vigour. Most weed seeds germinate within the top 5cm (2in) of soil, so adding 5cm of mulch on top of the soil makes it hard for some seeds to germinate and reach the surface.
  Mulches can be effective in lowering the weed seed germination rate but unless these are regularly topped up or replaced, their effectiveness decreases and weeds will eventually grow through a thin layer of mulch.

- Practice good hygiene. Clean tools (secateurs, pruning saws) and machinery to stop the spread of diseases.

- Use crop rotations to reduce pest pressure. Crop rotation might be a problem for broadacre farms as these tend to be monocultures of specific crops, but the principle is simple: Don't plant the same crop in the same paddock year after year. Mix up the crops to disrupt pest cycles and to repair the soil. To be really effective any crop might have anywhere from four to seven years rotation and this may be an issue for those farmers who tend to grow only a few different crops. But the problems are not insurmountable and the logistics can be determined.

  How effective crop rotation can be depends on when various crops are planted (the season), their different growth habits including the extent of root spread and above-ground leaf mass, duration of plant growth to maturity, and any specific cultivation or harvesting requirements.

  Crop rotation should be reviewed to ensure that it is effective as some parts of the farm may only support certain crops.

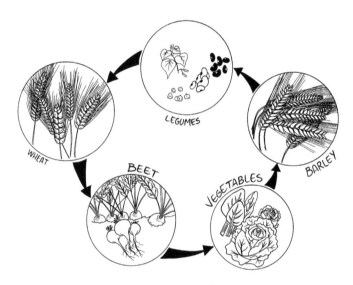

Figure 8.5 The concept of crop rotation.

- Sprinkle diatomaceous earth. This is spread over fruit and vegetable crops and the silica needles puncture the soft larvae skin of pests, causing their dehydration and death.

- Use livestock and poultry to eat weeds.

- Monitor soil fertility and characteristics – some weeds prefer acid soils, others low calcium. If the pH is far from optimum then this may be the overriding cause of some weed infestations.

- Provide soil cover – crops, cover crops or pasture. Species choices are crucial, and this is discussed in Chapter 5 (p.85).

Mechanical (Physical) Controls

Mechanical or physical controls are those things that are barriers or structures, as well as the use of machinery and tools.

- Use of insect traps and behavioural chemicals (lures) – including fruit fly baits, beer for slugs and snails. Traps can be sticky, lures, pitfall (beer in buried jar) or hiding places – rolled up newspaper, upturned wooden box, half an orange peel shell and loose hessian bands around trunks (removed with hiding bugs inside).

- Mechanical barriers, such as sticky and wet bases or bands (Vaseline or grease) around fruit trees to discourage climbing insects, bags over fruit (for insects), netting or other material to exclude pests such as birds from grapes, plant guards – for rabbits, rats and mice, and sawdust or wood ash around garden beds to discourage slugs and snails.

- Use tillage to disrupt breeding and expose insects to predators. To till or not to till? Remember that excessive tillage can result in soil erosion, breakdown of soil structure, a shift in microbial activity and loss of organic matter, and it uses considerable amounts of fuel and tractor time. On the other hand, the introduction of air into the soil is also important, especially in an organic system that relies on microbial activity to provide soil fertility.

  Not all soils, not all crops and not all farms are well suited to no-till. Many crops, such as small grains, clover and grass hay, can be successfully planted in untilled or lightly tilled soil. Ideally, using plant roots and animals like earthworms are used to actively till the soil instead of machinery.

  Some initial cultivation may be required to improve soil structure and breakup hard clay pans (using a Yeoman's plough for example), but continuous cultivation will adversely affect soil structure in the longer term. Ploughing, therefore, is not always essential and certainly ploughing that inverts the soil should be avoided.

  Remember that where no-till techniques are used, subsequent mechanical weed control options are more limited because the soil may not be loose enough.

- Practice cabbage moth (white butterfly) tennis, or use a fly swat.

- Install a UV bug zapper. This is most suitable for the backyard vegetable patch.

- Install bat and owl boxes in remnant pockets of bush and forest to help manage pests.

- Use attractants/food to induce predators into the garden – including water sources, blossoms and nesting boxes.

Figure 8.6 (L) A roller crimping attachment. (R) A Row mower attachment.

- Mowers and other cutting tools, rollers, roll crimpers, rotovators, spoon weeders, rotary hoes – for cutting, slashing, tickling, and the use of flame and steam.

  Flame weeders may kill broadleaf weeds (dicots) but may have little long-term effect on grasses (monocots) so reliance of these mechanical techniques may result in specific weeds dominating.

  The Lely weeder and the Einbock tine weeder are similar machines that have rows of tines (tynes) that tickle the ground and dislodge weeds. The spring-loaded tines either cause the weeds to be buried or lay on the surface. Other machines use an undercutter bar to slice through the soil, cutting roots off plants.

  For vegetables and other annual crops, where production practices keep natural plant succession at its earliest stages, the emergence of pioneer plants can become agricultural weeds.

  Mowing or rolling a cover crop to form an in-situ mulch can enhance the soil benefits of the cover crop, when compared to tilling it in, and can effectively suppress many annual weeds.

  Roller crimping is the flattening of weeds to control them and allow the organic matter to stay on the soil surface. The roller attachment is also used to knock down a cover crop so that a cash crop can be sown and grown amongst the decaying organic matter.

  Many weeding machines are used as a post planting technique and permit weeding between rows of crops. A row mower is another tractor attachment that simply cuts the plants in much the same way as a domestic lawn mower. It can move around trees and fence posts and can be useful in tight spaces.

  Something like a strip till drill or a no-till drill can be used to sow seed. Once the cash crop has been harvested and removed another cover crop can be sown if the original plants don't spring back and recover.

- Solarisation – plastic sheeting covering the soil to kill persistent weeds and steam weeding can cause some issues, such as loss of organic matter and nutrients, as the soil is heated. Heating the soil may also seriously affect beneficial fungi and bacteria and although these might eventually recover there may be some initial setback to production. Intensive heat applied to weeds causes water in the leaves to heat up and burst the cells as it expands. Steam weeders and flame throwers have their limitations – not all weeds will be successfully destroyed and you do need several applications over the year to better manage excessive weed growth.

Biological Controls

Biological control is crucial to pest management. This involves using natural predators to keep pests and diseases in check. Predators are not always pre-existing and present, so you need to encourage them to come. This may involve planting particular shrubs and trees to attract hoverflies, predatory wasps, lizards, birds as well as getting the soil right to ensure predatory nematodes and other soil biota thrive.

Numerous benefits arise from utilising natural enemies and other non-chemical techniques. Some are obvious while others are more hidden and difficult to quantify. In essence, natural enemies can enable control of pests in crops sensitive to chemicals as well as assisting in control of pests that have developed high resistance to chemicals.

There are two types of biological control agents – the predators and parasites. Predators include ladybirds, lacewings and praying mantis. They feed on their prey to complete their development.

Parasites include tiny wasps and some flies, and they are excellent biological control agents because they lay eggs onto or into specific pests. When the larvae emerge they eat their host and use it to continue their development. Many wasps are very small (just a few millimetres) but can lay hundreds of eggs inside pests and the whole wasp lifecycle might only take a few weeks.

- Encourage beneficial insects, animals (chickens, lizards and frogs). For example, frogs for caterpillars and predatory wasps for small insects. Poultry, such as geese, ducks and chickens, typically feed on grass and insects so can be used effectively to mow the grass weeds between rows of fruiting trees, and they will snack on snails and slugs.

- Specific biological pest control, such as fungus or bacteria to kill pests. For example, Dipel is a bacterium (*Bacillus thuringiensis*) that only kills caterpillars.

- Pheromones. These are natural chemicals that are given off to attract mates. They are used in traps, often including simple attractants such as yeasts and Vegemite® baits, to attract and trap fruit fly and other flying insects.

- Sterile insect releases. Sterile males are bred and released. These might successfully compete with normal males for mates and resources, so overall numbers slowly reduce.

The disadvantage of biological control and IPM is that it requires a greater understanding of the interactions between pests and beneficials, as well as the effects of chemicals. Some difficulties encountered moving to IPM include:

- Regular monitoring is necessary to identify pest outbreaks and their location within a crop. It may take time to develop suitable procedures and routines.

- Some damage from pests may need to be tolerated. Some pests may be required to support a useful population of the natural enemy.

- Good timing is necessary when introducing natural enemies that are bought and released – not too early, not too late. It is important to get natural enemies established quickly.

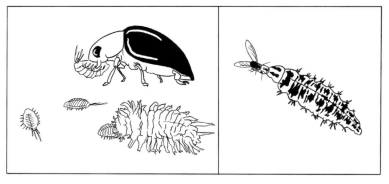

Figure 8.7 (L) Adult and larva *Cryptolaemus* beetles (a ladybird) feeding on mealy bugs. (R) Green lacewing larva eating a winged aphid.

Chemical Controls

Chemical controls involve the use of natural and artificial (human-made) substances which are typically poisons.

- Use organic sprays. Yes, there are a number of organic (even certified organic) weed and pest treatments available. Some of these are based on pine oil, and others on vinegar and salt. Always read the label to make sure the product is suitable for your needs.

- Included here is pelargonic acid, also called nonanoic acid (one commercial brand is Slasher Organic Weedkiller). The acid is named after the pelargonium plant, since oil from its leaves contains esters of the acid. The ammonium salt form of nonanoic acid is used as a herbicide. It works by stripping the waxy cuticle of the plant, causing cell disruption, cell leakage, and death by desiccation.

- Use plant-derived sprays to control or repel insects and other pests, fungi and viruses. For example, pyrethrum, wormwood, nasturtium, horsetail, garlic, chamomile, neem oil, citronella and lavender oil. Most of these repellents are made using water, but some may involve an organic solvent such as methylated spirits.

- Natural chemical control. For example sulphur, tea tree oil, derris dust.

- Insect growth regulators and hormone disruptors. These disrupt the pest growth cycle.

- Oil and soap sprays (Natrasoap) – for aphids and scale.

- Pesticides (commercial) – using the most appropriate product at the best time and after careful monitoring of pest levels. Check toxicity and nature of the chemicals – some plant-based pesticides are available that contain pyrethrum.

Whenever possible, spray at night to ensure no indiscriminate effects on bees, butterflies and other beneficial insects.

Final Remarks

Integrated pest management (IPM) is an effective and environmentally sensitive approach to pest management that relies on a combination of common-sense practices. IPM programmes manage pest damage by the most economical means, and with the least possible hazard to people, property and the environment.

Pest and weed management on transitioning farms require a holistic approach, as no two years on a farm are ever the same. It relies primarily on preventing and avoiding pests with cultural and mechanical suppression, rather than using chemicals as the first option.

Regenerative agriculture is not anti-pesticide but rather a move towards a diverse range of techniques and strategies to build a functioning farm ecosystem. In time, chemical use will correspondingly decline as weeds are seen as animal fodder, diverse cover crops reduce insect and pest attack and increasing soil health builds resilience. Regenerative agriculture is about seeing opportunities rather than obstacles to overcome.

9

The Transition to Regenerative Agriculture

Regenerative agriculture is an umbrella term for any agricultural practices that lead to building more carbon in soils and thus healthier soils, increasing bio-diversity and hydrology in the farming landscape, and still making all of this economically viable and profitable.

All agricultural farms are businesses, and they exist to make money for the farmers. Various enterprises are undertaken to generate income and cash flow, and these enterprises can change from one year to the next. Farmers need to make a living and they can do so by incorporating a range of regenerative and restorative techniques.

By now you have realised that we need a lot of tools and skills to help us with the regenerative work on the soils we have. Farming is a complicated process, and many people have misconceptions about it and don't appreciate how difficult it can be.

Farming is diverse. It ranges from low numbers of cattle grazing on vast areas of native grasses in rangelands, to cereal and stock rotations in the 'wheatbelt', to intense animal feedlots and battery hens.

I would like to think that most farmers are conservationists – they live on the land and derive their livelihood from it. They need to manage their properties well in order to make farming a viable business. So it is in their best interests to care for the land. Even so, some farmers, who have inherited European ways to farm our land (e.g. ploughing to turn over the soil), also have little understanding of how to farm in a land-healing way.

Farmland throughout the world is increasingly becoming subject to dryland salinity, soil acidification and loss of soil carbon. These all affect productivity and ultimately the sustainability of the farming enterprises. The options for good land management systems that support sustainability are well known, but in some cases are not economically viable or the most favourable, and this needs to be addressed.

Farming is a risky business, and in our day and age, when we are all about mitigating risk, farmers need to have a good understanding of what the market-place wants. They might believe in planting particular tree crops or timber but if there is no market for that product or no timber mill to slice that timber for sale, failure is inevitable.

Sometimes it is hard to know if the path we are walking on is taking us in the right direction. It seems we often stray from the path or find side paths which we follow until we reach the dead end, turn around and endeavour to walk the right one. If you are farming in particular ways and you aren't getting the results you want, it obviously isn't working. Stop doing it. You need to see farming through new eyes, develop new approaches on how you see the land, the soil, the animals, the nutrient cycles, the water cycle and even the crops you are growing. No-one likes change, but at the end of the day if things aren't working, stop doing them.

Shifting to regenerative agriculture requires a shift in the world view and paradigm we hold. We have to let go of the old 'we have always done it this way' to 'what can I do to radically change what is happening on the ground?'. It may be too hard for farmers to unlearn what they know and do, and then learn new ways of doing things. Regenerative agriculture and organic farming was all farmers knew for hundreds of years, right up to World War II (WWII). Now, 70 years later, we are going back to what our grandparents did because science says it is the best way to farm and care for the planet.

The principles of regenerative agriculture recognise that farms and properties are part of a larger ecosystem and our understanding of the interactions between soil, air, water and organisms allows farmers to establish cycles of regeneration. Whereas conventional agriculture is extractive, regenerative systems improve the land. It's like a bank account where you might have a lump sum to start with and then you just take money out (conventional agriculture) or you could still take money out but now and again you deposit into the account and your money doesn't just waste away but keeps earning you interest (that's regenerative agriculture).

Sometimes change comes after a crisis, like ravaging bushfires, extended periods of drought or destructive flooding, hail and storms or after extremes of weather. But why do we need to wait for a crisis before we change what we are doing? Shouldn't we all endeavour to assess what we currently do and move in other directions to make a brighter future for all? Humans have a very simple philosophy 'if it isn't broken, don't fix it' so we carry on struggling because we get by and we just don't want to make an effort to change.

Transitional change is transformative – it will require things to be done differently. Farmers, like all other people, are frightened to take that first step for their new journey. We all know from our own experiences that the journey itself will be alright and that they will cope with the challenges along the way. It's just that first jump into what they see as the unknown that is scary. It takes courage to change the direction of the ship you are steering.

Finding a mentor who has started their own new journey can be helpful – if only for them to be a sounding board or devil's advocate to your ideas, concerns, dreams and ideals. There is a collective wisdom out there willing to share and help others to achieve their goals. You are not alone. Later, you can help and support others to start their own journey.

Thinking outside of the box can be difficult for some, but this is essential for real change. We shouldn't be driven by fear but by hope for a promising future.

REGENERATIVE AGRICULTURE SHIFTS THE PARADIGM

COMPETE WITH NATURE

DISTURB SOIL

MONOCULTURE

SIMPLE ANALYSIS

LITTLE ENVIRONMENTAL CONSIDERATION

WORK WITH NATURE

COVER AND PROTECT SOIL

DIVERSITY

HOLISTIC

ENVIRONMENTAL PROTECTION

Figure 9.1 Regenerative agriculture requires a mindshift and a different world view.

It is the fear of bankruptcy and losing the farm that frightens people. They are frightened to fail. But change is always possible and doable, you just need to believe in yourself and your ability to make it work.

Fear of the unknown, fear of failure, fear of acceptance, fear of judgement by others, fear of ridicule from other farming neighbours and fear of market uncertainty limits a typical farmer from taking that leap, almost a leap of faith, to start a new journey and forge a new path. Other farmers may see your efforts as a threat to their way of life, of their ideals and farming technologies but all of this is really unfounded. Too many regenerative agriculture farmers are finding great success.

These initial perceived hurdles stop many farmers from taking that first step. Then we must overcome the fear of doubt: Doubt that regenerative agriculture will not work in their area because it is too dry, too wet, rocky or hilly, has too much seasonal snow, and every other excuse that casts doubt into the farmer's mind.

No-one likes change – it's just too unknown to do something different, to go out on a limb and to do something new. People have this (unnecessary) fear of failure, of maybe over-estimating the possible hardship and they may lack confidence in their own ability to have a go and make it happen. You need to break the cycle of moving from one disaster to the next. While everyone hates change, it is needed if we are to repair damaged landscapes and farms responsibly.

Some types of change we resist, and at times we just can't predict which direction change is coming from and where it might lead to. It's the unknown that creates concern, and this doesn't have to be anxious and terrifying, but exciting.

The transition to regenerative agriculture is a paradigm shift, a shift in thinking about how land is used, food is grown and ecological literacy. Ecological literacy (or ecoliteracy) describes our understanding about living systems (ecosystems) and nature which make all life possible. It challenges our world view and what we hold as true.

Farmers have to shift the mentality that this is all we know about it, that this is all we can grow, and that this is all we can do. Once that mindshift occurs then all things are possible. Farmers need to let go of the desire to maintain control. At this stage, farmers have that lightbulb moment when they realise that all they believed in and did was wrong. The transition to regenerative agriculture doesn't have to be overwhelming, even though a farmer is really moving in the opposite direction to what they have been doing for most of their life.

Some Steps to Transition

The transition to regenerative agriculture is not a destination, but a journey. There is no finish line. The transition steps will vary from farmer to farmer. Be informed and start by finding out about farming practices that are making a difference. Attend workshops, field days, read and research. Measure what you have by various soil tests that indicate not only nutrients, but the biology, carbon: nitrogen ratios and so on. This is the benchmark which can be used to measure your future success.

The transition to regenerative agriculture is not without hurdles. From the graph below there may be some initial drop in income and profit as the system, techniques and operation is changing. Once the soil begins to improve, plants flourish and income rises from the cost of lower inputs and hopefully greater return on quality produce, the profit will slowly increase.

Regenerative agriculture does not focus on refining techniques or substituting inputs, but rather a complete shift in the ways of farming. The mindset needs a shift towards a commitment to see it through – maybe a commitment to do one or two paddocks for a few years. It has been my experience that once a farmer does make that change, they don't go back. Sure, there are ups and downs, challenging things get thrown your way, but hasn't that always been the case? You learn to adjust, to reassess and digress, and then move forward again.

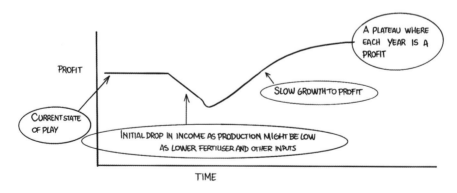

Figure 9.2 Changes to profit during the transition phase.

The first step is to have a willingness to change from conventional farming practices and adopt new methods to deal with challenges as they arise. The second step is to inform yourself, using good science and creative thinking to come up with solutions for your property. Along the way you should become aware of your responsibility to protect the resources that are available to you, and that includes the air, soil and water that passes through your property. At the end of the journey I know that you will be leaving the land fit for a future generation of farmers.

If we are asking farmers to make the transition then they must be provided with appropriate tools and resources, whether that is training, small grants, expert advice and consultancy, mentoring or other government or industry support.

The shift to regenerative agriculture must coincide with education and training. Once recognised and accredited courses are rolled out, then there will be opportunities for young farmers to become better informed and then to make that change.

There is no specific toolkit to transition nor a universal step-by-step procedure. However, the following points are worth considering:

1. Set goals that are realistic and achievable. Have a five-year plan. Develop a whole-farm plan, a drawing that maps out what you intend to do. Design for resilience with the aim (ideally) of no soluble fertilisers and no pesticides and fungicides.

2. Not all at once. Take it slowly. Transition one paddock at a time if that is all you want to start with. You need to see what happens, make adjustments, if required, and then add another paddock.

3. Leave the difficult, most depleted and most degraded paddocks or areas to last. Start with those areas that are in fair condition and with a little effort and expense can bounce back and be more productive with better soil fertility and structure. Start with things that have the lowest cost and the lowest risk.

4. Aim to reduce fertiliser and pesticide use in that paddock or better still over the whole farm. For transitioning farmers, be aware that before you stop adding artificial fertilisers altogether the soil needs to be in a state where it has the ability to restore and improve. Once soil colour and structure starts to change then you slowly reduce chemical fertiliser use. Along the way you would be planting cover crops, maybe adding soil amendments to change the pH, the rate of infiltration or specific mineral deficiencies. You may notice the fine threads of mycorrhizal hyphae increasing in the soil, and the soil colour becoming darker as more organic matter accumulates and soil aggregates develop.

5. Focus on soil structure and soil health which also includes organic matter, water availability and permeability. Transition can be quite quick if there is reasonable organic matter in the soil already.

6. Focus on soil life, especially beneficial bacteria and fungi. Focus on biodiversity, ground cover, no till, soil fertility.

   Really focus on the soil. This is the key. Use amendments to change nutrient levels and increase beneficial fungi and bacteria. The nitrogen, phosphorus and other nutrient cycles work to improve plant and animal health. Continue to enhance and maintain the landscape and the environment.

7. Focus on profit not turnover. It should be more about how much money is made and not how many tonnes you get per hectare. Farms needs to be both regenerative and profitable. What is the cost of transition? This is largely unknown but when you consider the reduction in the cost of inputs then farmers should be able to see that they will get a return. However, until the soil health and fertility is better, the profit may not be too great, but hopefully this increases in the years that follow.

8. Develop a flexible management plan. Monitor, measure, reflect, change, maintain records, make notes, develop a strategic plan, implement the design, and be able to adapt and change the plan as required.

Finally, transitioning to organic is a long process. Most organic certifying organisations have a three-year period of conversion where farmers do not use chemicals but cannot promote their produce as organic and receive premium prices. At most they may be able to sell produce as 'chemical-free'. Farmers can expect low outputs and low income while their soils are slowly rehabilitating. There are also costs associated with certification fees and soil testing that are essentially non-refundable. Farmers might just farm using organic principles, but not go through the certification process. That is a decision you must make.

Measuring Success

Continual monitoring is essential if you want to determine if things are getting better or worse. Some equipment, such as a rain gauge, max-min thermometer, pH test kit and a brix refractometer are useful and a relatively cheap investment. These allow you to measure some basic things about the climate, soil changes and nutrient density of plants. A salinity (TDS) meter is also useful, or for very salty dams and landscapes a refractometer that measures the percentage salinity is essential. Both a compost thermometer, which is used to determine soil temperature, and a moisture meter (cheap ones are not reliable but scientific ones can be expensive) are useful in the field.

We need to measure so we can adequately manage the system, but the aim is to measure to improve the system. When we measure the increase in biodiversity we measure the profitability, and it is confirmation that we are doing something right.

As a simple guide, check out this example of an acrostic for MONITOR:

Manage: Use the measurements and observations to guide your responses.

Observation: Make written notes of what you see.

Numerical data is collected and recorded.

Inform: Use the data to inform management and the strategies to adopt.

Track changes of readings. Predict future trends.

Organise a schedule to make regular observations.

Reflect: How is it all going? Is what I am doing making a difference? What do I need to change?

Measuring and monitoring is the best way to gauge your successes and failures, and to forge new paths and directions if that is required.

What Kind of Crops and Enterprises

Modern agriculture is a right livelihood occupation. It can bring personal fulfilment, increasing your sense of worth and can be a good way to make a living. Many practitioners feel a connection to country, to the land, but many current agricultural practices are causing contracting communities. Since WWII there has been a slow progression of reduced diversity, declining profitability, decreased nutrients and increased inputs. Alongside this, human health is deteriorating and the levels of minerals in food has been steadily declining since this time too. Besides poor quality food, we now have poor quality soil and degraded ecosystems.

What we eat, how it is produced and where it comes from can really make a difference. The saying 'we are what we eat' is actually true – much of the current evidence of obesity, heart disease, liver failure and cancer is simply due to our lifestyle choices and a large part of that is our diet and the foods we eat. With more farmers turning to regenerative agriculture it is hoped that food quality will improve.

To support this trend there needs to be more connection between the farmer or producer and the consumer – more dialogue and more interaction. Consumers who buy produce solely from the grocery shop are divorced from all aspects of food production, while the farmer who doesn't know who eats his products just farms to produce a yield. There can also be a difference in what people understand about what farming is and what a farmer is. Do we see their 'occupation' as a steward of the soil and environment or more simply as a grower or producer of food, textiles and other produce? Do farmers themselves see their farms as a business or a way of life or somewhere in between?

On page 172, we mentioned that farmers need a mindshift. Similarly, consumers also need a mindshift – to eat only locally produced food, to only eat seasonal food and to endeavour to support local farmers and growers. The community, generally, needs to value farmers and support them as best they can. More customers are seeking demonstrable proof that the food they are buying is produced in ways that haven't harmed the environment. Essentially, your customers want to know how their food was grown.

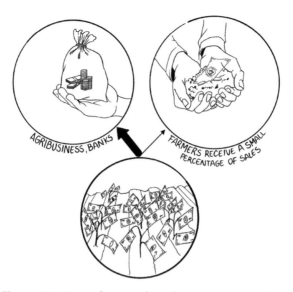

AGRIBUSINESS, BANKS

FARMERS RECEIVE A SMALL PERCENTAGE OF SALES

Figure 9.3 Most of money from farming goes to others.

Some new generation farmers sell direct to customers or develop co-operatives with other nearby farming families and produce food for local needs. It is not always possible for farmers to sell (more) directly to the local community so they are forced to sell to large organisations. This might be desirable anyway if farmers have long days and then have to take produce to the weekly farmers market which could still be some distance away. Marketing and promoting your product or brand can also be costly and time consuming. Furthermore, the lack of local or regional processing facilities is a major sticking point for many communities working toward a local sustainable food system. The dilemma for the farmer becomes: Do I make more money selling direct (with more work) or sell cheaper to agribusiness for less work?

Unfortunately, the most money made from farming is made by people on the other side of the farm gate – those involved in transport, distribution and the grocery stores. Food policies are serving global organisations and markets and not really serving the people. Wealth is not with farmers, but multinational companies and agribusiness who are making large amounts of money while farmers are going broke.

Farmers have emotional attachment to the crops they have been growing, but, at the end of the day, they need to grow what they can sell. However, consumers actually have little impact on the marketplace. Agribusiness shapes the markets. Farmers grow to satisfy the demands of chemical companies, machinery dealers, seed merchants and all of the other companies involved in the food supply and distribution chain.

Farming is a commodity-driven system where there is profit for agribusiness, but little profit for farmers. Farmers are loaded with debt, and we need to look at the bigger picture and take a more holistic view of all potential outputs from a farm. Too often, larger farms become specialised and their success and failure depends on market forces, so we need to work on ways to change this.

Profit is always a stronger driving force than any social or environmental concerns. Large companies only change when it is profitable to do so. Most farmers will not change simply due to climate change concerns, consumer health, animal welfare, soil degradation or other environmental issues. As we have already mentioned, farmers, like everyone else, often won't rush in and adopt change easily.

Profits are determined by the cost of all the inputs, price at the marketplace and turnover (number or weight of goods sold). This allows the overall profit and margins to be calculated for the product or per unit of product – if that's a cow or a tonne of wheat.

When the cost of inputs increases the farmer may not be able to get a higher commodity price. This translates to lower net income, and the only choice a farmer may have is to pay less for labour or to reduce other expenses.

As input costs (pesticides, fertilisers, grain) continue to rise year after year the profit margins typically fall as the return on products declines. Debt for farmers steadily rises as they pour money into larger machinery and more fertilisers while receiving less for crops, wool or stock. The farmer basically gets less money than the grocery store and butcher as well as those who do the transporting, slaughtering and packaging. There is a horrible imbalance of what the farmer receives and what everyone else gets in the supply chain.

Furthermore, there is a vast difference between productivity and profitability. Fertiliser is expensive and farmers have been heavily dependent on NPK fertilisers for quite some time. Conventional farmers may add fertiliser as a knee jerk reaction, boosting nitrogen or phosphorus in the soil stock, whereas regenerative agriculture farmers may look at the issue in a holistic way, examining soil structure, the soil life (especially nitrogen-fixing bacteria, mycorrhizal fungi and hosts of others), the amount of carbon in soil (organic matter) and examine soil composition in light of waterlogging or other issues. Conventional farms may find increased production where profitability falls, whereas regenerative agriculture farms may find lower productivity but higher profitability, especially if crops are grown organically.

Diversifying what is undertaken can burden the farmer with more work, but once systems are in place it is just about management. Layering a number of farm enterprises in the same area is a form of stacking so you could imagine crops, cattle and then chickens all following each other in a paddock – a sort of stacking in time. Let's start the process of working smarter, not harder.

Again, animals typically play a key role in every ecosystem, so while you can have a plant-based regenerative farm it seems that those farms that incorporate animals have several advantages. Many animals and plants thrive and survive in low input systems, and have done so for millennia. You just have to look on your property for plants that defy existence – it hasn't rained for ages, the soil is poor, hard and dry, and yet these plants are alive.

Whatever you decide to grow and produce there are still some issues to address. Genetically modified (GM) crops initially showed promise and hope.

While humans have been manipulating the genetic makeup of plants and animals for centuries, producing new varieties of apples, crops that are disease resistant or animals with higher meat quality, recent endeavours involve the transfer of specific genes from one organism to another. These genes might enable a plant to resist insect attack, produce better quality oil for cooking or be able to better withstand drought and low rainfall.

Unfortunately, it just didn't stop there. Some companies produce GM seeds to be resistant to glyphosate and other herbicides so these can be used to kill weeds but leave the crop alone. However, it soon became apparent that insects, disease and other pests became resistant to glyphosate and other chemicals as well. Genetically modified organisms (GMO) are not accepted by organic farmers.

Different countries have different policies on GM foods. For example, it is believed that about 80% of processed foods in the USA contain GM ingredients, yet the foodstuffs do not need to be labelled as such. In Australia and New Zealand, GM foods are more regulated and labelling is mandatory.

Additionally, many farmers are forced to pick fruit and vegetables before they are ripe to ensure they won't spoil during the long food miles transporting produce to the marketplace. Often this means the food products are not reaching full nutritional and taste potential. So this is another good reason to supply local needs.

Finally, ethanol production is not sustainable. It seems a great move forward to be making biofuel but it relies mainly on corn. Nearly half of the corn produced is used to make ethanol, nearly half is used to feed animals and only a small amount (10-15%) is used by humans. As more renewable energy comes to the fore, ethanol production and demand will decrease and farmers who have planted almost all of their farm to monoculture corn will financially suffer. Farmers really do need to think all these things through.

Family Transition

Farming can be a seven day a week operation. Some farmers work every day and some of those days are long – out doing chores from sun-up to sunset. This is hard for one family farm where family (and community) life suffers. A large farm might really need several families or extended families to operate it.

While aging farmers would like their children to take over the farm, this is not happening for a few reasons. Firstly, there is just not enough income for more than one family to run the operation. Many young people come to the city for other types of jobs, and young people know from their own experience that farming is a thankless business: You work every day doing something to maintain the infrastructure with little recognition.

Many farms are seen as family-orientated businesses, but often there is no clear exit strategy for parents to hand the farm over to one or more of their children. Older farmers may wish to retire and get some money for the farm. Their children may not be financially able to buy it or even contemplate paying it off over time, so the farm is often sold to others.

Complicating any family handover is that some young people experience disempowerment as they cannot seem to make their voice heard, often having a different opinion to that of their parents. Young farmers may want to try different crops and strategies but dad is old school and struggles to accept anything different. Then again, some children hold the same views as their parents, so it becomes, how do you break the habits developed over a lifetime and, in some cases, several generations? 'My father did this and his father' and so on. Old habits die hard.

Some farmers also have different opinions to the scientific community. For example, scientists and researchers rate biodiversity and conservation measures as more important to agricultural production, ecosystem resilience and sustainability than farmers. Farmers tend to value information from the government and agricultural sector to make decisions. It's like farmers have shutters over their eyes as they are so busy working on making a profit and chasing higher yields. Thankfully the tide is turning in farming communities but the wave is gentle and, like a tide moving into a river, it is slow and steady.

A major barrier to convincing more farmers to follow the regenerative path is grounded in whether they can compete with conventional agriculture when it comes to yield and profits. Yield has long been held as the holy grail of farming, but maybe we should look more at money in the bank. Regenerative agriculture can allow farmers to get out of debt and obtain a profit advantage.

Up to this point when farm profits are minimal each year a farmer needs to make a choice – get bigger or get out. So farmers are buying up neighbouring farms, buying larger machinery to work these larger combined areas, but still they just make ends meet. While few in number, large-scale farms are generally industrial but they produce the majority of foodstuff we consume.

"Get big or get out" still haunts farmers since it was first voiced in the mid-1970s by Earl Butz, USA Secretary of Agriculture. It is almost like saying "buy out your neighbour or lose your own farm". Many farmers still view this statement as a threat, even though getting bigger does not ensure proportionally greater profit, reduced risk or job and business security. Often the farmer just gets saddled with debt. This issue then becomes a problem for the younger generation who are trying to get into the farming business or to buy the farm from mum and dad.

Getting bigger means fewer farmers, lower numbers in the local community, less money spent in the local community as larger companies tend to buy cheaper inputs elsewhere, and the result is dying towns. This can have a cascading effect, shops start to close, people leave, the school shuts down and the small number of students are bused to larger towns nearby.

Diversifying is the road to expanding a farming operation. You don't need to become bigger to enjoy your work more and to have enough income to meet the family's needs. You just need to farm differently. Farmers can defy the logic of get big or get out by transforming their properties. Nature's tools are free and at our disposal.

Many of the new breed of farmers taking on regenerative agriculture, who are in every way ordinary people, are doing extraordinary things. Too often farmers do not get the recognition they deserve – gratitude for producing food, an acknowledgement they are doing good and that their efforts are valued by the community. This acknowledgement by the community needs to happen too.

Final Remarks

It is a sad reflection of the farming community when producers simply follow the advice of conventional agronomists and chemical representatives who tell them what, and how much, to add. Regenerative agriculture challenges farmers to take more (full) responsibility for what they are doing to the soil and to organisms and make informed decisions about the effects of their actions on the community and the environment. For example, it is not sustainable if we have to feed stock with plants requiring lots of fertilisers and import food from elsewhere. We should now recognise that the soil biology supplies the plants with the nutrients that farmers are currently buying from outside the farm.

In some circles, regenerative agriculture is promoted as our saviour to fight climate change and many governments and corporate bodies are funding regenerative agriculture projects. Some countries have carbon farming schemes which generate carbon credit units. This is where a long-term storage contract is negotiated with government or some organisation, maybe for 25 years. As an example, this may equate to 15-25 Australian dollars a unit but it is not fixed. One unit is usually 1 tonne of carbon dioxide which is about 270kg (42.5 stone) of carbon.

Farmers normally register a project where they will undertake a range of activities such as tree planting, adding compost, biochar or other organic matter substances, stubble retention, no-till cropping or some other pioneering enterprise. Carbon credits is one way to add value to areas of the farm that may be degraded or not as arable as other parts. Some schemes may allow adding gypsum or other amendments to improve soil structure and composition. There are usually requirements to measure the levels of carbon in soils over many years. Maybe we need other incentives to increase the uptake of efficiency measures or better farming practices. Whatever happens in the future, farmers need to feel that they are part of the solution.

While there is ample proof that regenerative practices can sequester carbon, it is also clear that if we have any intention of mitigating climate change we would need every farm all over the world to jump on board. For small family farmers such a transition can be costly. While I would like to think that regenerative agriculture will significantly reduce the effects of climate change, I do not believe that it is the one and only solution to this worldwide problem. Certainly regenerative agriculture should not be touted as 'the answer', but just part of the solution.

Finally, agrichemical business is all about the 'war with nature' whereas we should be 'working with nature'. Trying to fight nature is a battle we will never win. Nature is more complex and sophisticated than we acknowledge and she is a great teacher. We just need to be better students.

Index

acidification 29-30, 55, 84, 111, 169
acrostic 12, 174
aerobic microbes 133
agrichemical 180
agroecology 1, 9
agroforestry 5, 102-4, 136, 150
air quality 3, 24, 114
allelopathy 56, 157
alleycropping 105, 109
amelioration 5, 57
ammonium 30, 36, 44-5, 78-9, 167
anaerobic digestion 132
Andrews, Peter 130
anoxic 34
antagonism 74-5, 143, 145
anthropocentric 2
arbuscular mycorrhiza fungi (AMF) 53-4, 181
azospirillum 80

beneficial insects 24, 113, 153, 155, 158, 161, 166, 168
biochar 60-1
biodiversity 4, 7-8, 11-12, 17, 21, 24, 27, 52, 64, 96, 105-6, 110-1, 114, 118, 123, 136, 148, 153, 158, 169, 174, 179
biodynamics 112, 115
biofertiliser 35, 132-4
biopesticides 155
biostimulants 62, 131
brittle 28-9, 127
Brix 67-8
buffer strips 8, 107, 162
bushland 3, 20, 114
Butz, Earl 4, 179

CAFO 138
carbon credit 180
carbon dioxide 25, 40-1, 43-4, 50-1, 57, 61, 65-6, 69-70, 72-3, 78, 132, 136, 154, 180
carbon drawdown 7
carbon farming 180
carbon sequestration 6, 20, 44, 63, 106, 137, 153
cash crop 70, 98, 100-1, 112, 125-6, 165

cation exchange capacity 30, 100
cell grazing 116, 120-1
chaparral 117
climate change 6-7, 15, 18, 42-3, 119, 134, 151, 177, 180
climate uncertainty 6
Cluff, Darryl 125
colloid 36-7
compaction 19, 27, 34, 49, 57, 84, 92, 140-1
components 8, 15-7, 47, 66, 125, 138, 142, 146-7
compost teas 134
conservation agriculture 1
consumer 136, 175, 177
contour maps 15
conventional 1-4, 6, 9, 18, 20, 111, 113-5, 117, 125-6, 153, 155, 170, 173, 179-80
crop-livestock 147-8
crop residue 147, 162

deficiency 75, 142
denitrification 45
dispersion 38-9
DNA 44, 66, 73, 80, 82, 142
Doherty, Darren 128

earthworm 51, 60, 113, 133
ecocide 111
ecoliteracy 9, 171
ecological 2, 6, 9, 14, 85-6, 102, 115, 131, 148, 151, 153, 155, 171
ecosystems iv, 6, 9, 15, 21, 87, 106, 108-9, 111, 117-8, 129-30, 135, 140, 143, 153, 171, 175
ecosystem services 27, 63, 89, 106-7, 119-20, 131
endophytes 82
environmental degradation 1
epiphytes 88
erosion 2, 8-9, 19, 22, 27, 42, 45, 49, 57, 84-5, 89, 94, 96-7, 99-100, 107, 111, 114, 119, 127, 130-1, 148, 164
ethanol 9, 66, 178

factory farming 135

farmer 1-3, 9, 11, 23, 26-8, 63, 67, 85, 94, 98, 102, 120, 125, 127, 134, 144, 171-2, 175-7, 179
fertilisers 1-3, 9, 21, 27, 30, 36-7, 40, 45-6, 54-5, 58, 64, 75-7, 91, 112-3, 117, 125, 131, 134, 136, 141, 144, 155, 173, 177, 180
fodder 81, 94-5
food miles 18, 25, 114, 178
food production 1, 3-4, 8-9, 26, 45, 115, 123, 152, 175
food sovereignty 11
forage 69, 94-5
Fukuoka, Masanobu 89-90
fungal 35, 49, 51-2, 75, 88
fungicides 156

genetically modified (GM) 153
genetic manipulation 2
Gordon, Ethan and Lorraine 13
grazing 5, 7, 13, 49, 95, 97, 103-4, 112, 116-8, 120-3, 125-6, 135-41, 144, 146, 151, 169
greenhouse gas emissions 6, 25, 109, 135, 136
Green Revolution 1
guilds 12, 90, 124, 157
gypsum 38-40, 58, 180

Haggerty, Ian and Dianne 134
heavy metals 47, 55, 76, 84, 143, 145
Heenan, Lisa 128
herbicides 1, 3, 6, 11, 20, 51, 62, 74, 111-2, 114, 117, 125-6, 134, 153, 155, 158-9, 178
herbivores 91, 118-20, 123, 138
holistic management 5, 9, 112, 152
holistic planned grazing 7, 13
Holmgren, David 123-4
hot composting 157
humus 20, 36-8, 41, 47-8, 60, 133, 157